Herman Jackson

COMPLEMENTARITY IN BIOLOGY

COMPLEMENTARITY IN BIOLOGY

Quantization of Molecular Motion

James P. Isaacs and John C. Lamb

THE JOHNS HOPKINS PRESS : BALTIMORE

This work was initiated at The Johns Hopkins University School of Medicine, Baltimore, Maryland, under U.S. Air Force Contract AF49 (638)–590, and completed with the support of the Glancy Foundation, Atlanta, Georgia, where the authors have been Medical Scholars.

Standard Book Number 8018-1039-6

Manufactured in the United States of America

Library of Congress Catalog Card Number 69–16860

ACKNOWLEDGMENTS

We have been fully aware during the writing of this manuscript that the making of a book involves many more persons than the authors. The material, morale, and critical support of those involved is as important as the setting of words to paper. Financial aid for the initial phases of this work was provided by the U.S. Air Force Office of Scientific Research under contract number AF 49 (638)–590 to the Surgical Department, The Johns Hopkins University School of Medicine. The major portion of the research and the writing was accomplished while the authors were Medical Scholars of the Lenora and Alfred R. Glancy Foundation, Atlanta, Georgia. We are indebted for the freedom which characterized the assistance of these organizations. We acknowledge, also, the generosity of Mr. Theodore J. Carski, of the Carski Foundation, Baltimore, Maryland, in assistance with the publication cost.

The development and virtues of any book, we have found, depend in large measure on the evaluative "feedback" of those who read the manuscript in various stages of completion. An early and more compressed version of the manuscript was read by Dr. Thomas L. Lincoln of The Johns Hopkins University School of Medicine, and by Dr. J. M. Burgers of the Institute for Fluid Dynamics and Applied Mathematics, University of Maryland. Their comments sharpened our interest in the subject and led directly to the realization that considerable revision was needed. Dr. Irwin W. Sizer, Dean of the Graduate School of the Massachusetts Institute of Technology, read the first draft and offered encouragement concerning the basic ideas.

In somewhat its present form the manuscript was read by Dr. Augustus L. Stanford, Jr., Georgia Institute of Technology; Dr. Walter S. Koski, The Johns Hopkins University; Dr. Joyce J. Kaufman, Research Institute of Advanced Studies, Baltimore, Maryland; Dr. Victor A. McKusick, The Johns Hopkins University School of Medicine; Dr. J. M. Burgers; and Dr. Sizer. We thank these individuals for their generous expenditure of time and effort in giving us the critical review which was essential.

Dr. Warfield M. Firor, Professor Emeritus of Surgery at The Johns Hopkins University School of Medicine, occupies a unique position in the genesis of this book. It was he who provided in his gracious and gentlemanly manner both the goad and the rein at many times during its inception.

Of course we can place the blame for none of the faults of the book upon these persons, but we must acknowledge that they have contributed a great deal to that which may be acceptable.

No listing of acknowledgments could be complete without including our wives, Arlene and Marjorie, whose intuitive understanding of things aborning greatly eased the pains accompanying this book. We dedicate this book to them.

CONTENTS

List of Figures (*continued*)

List of Tables

COMPLEMENTARITY IN BIOLOGY

INTRODUCTION

That *all* living processes will eventually be shown to be mechanistically determinate is the main antithesis of this monograph. The following statement concerning mechanistic explication in biology is reproduced because of its conciseness and clarity and for the important reason that its author has made three oblique steps toward limitation of the use of mechanistic determinism in description of biological processes. Morowitz states:

> Thus in the ideal case one should be able to start out with a quantum mechanical description of the atoms making up a cell and by application of the principles of quantum mechanics predict the cell's behavior. Such a description of a living cell is far from that realizable at the present state of knowledge.
>
> Detailed atomic description of the structure of a living cell consists in giving the position of each atom and the bonds between neighboring atoms. The functional description of a living cell thus consists in following the time dependence of the detailed atomic description. In view of the complexity, [i.e.] number of atoms in a living cell, no one hopes to give as detailed description as this. Rather, one assumes that laws governing aggregates of atoms will simplify the description, and

> laws of cellular biology will emerge (Morowitz, 1960, Vol. 3, pp. 111–13).

Biological laws are here stated to arise by a straightforward deductive process from the quantum mechanical laws governing the behavior of atoms and molecules. Resort to intermediate laws pertaining to aggregates of these particles, i.e. physical and chemical generalizations, is somewhat a matter of convenience, since they can be deduced from the same quantum mechanical principles.

It may be instructive to view Morowitz's statement in a somewhat different light as an attempt to humanize Laplace's calculating demon. The concept of the demon in the 18th century gave the modern impetus to mechanism as a complete description of a physical process. In the original idealization, the demon was empowered with a knowledge of the simultaneous positions and momenta of all the particles of the universe along with Newton's laws and, hence, the ability to determine or calculate all subsequent positions and momenta. The demon in Morowitz's statement is unable to fulfill the criteria of determinism, i.e. is unable to break the limitation of indeterminacy which is incurred during interaction of measuring device and process being measured. This is to say that, as a first humanizing step, the demon is necessarily interpreted as possessing quantum mechanical laws and subsequent initial values relating to atoms and molecules instead of elemental particles. As a second step, he is also given an abstracting ability and some biological knowledge. The overlooking of this capability has been a criticism which Polanyi (1958) has raised with respect to calculating demons. This is to say that if the demon had human capabilities, he would work toward a proper couching of his data in terms of similarities and differences rather than endless collections of numbers. As a third step, Laplace's demon is also allowed a statistical averaging ability in terms of physical and chemical generalizations. This complies with the criticism of Born (1962, Chap. 1, 4, 5, and 8) and Brillouin (1962, Chap. 21) that the initial values of position and momentum of a purely mechanical system have an inherent spread which increases with both time and observation. A fourth step must now be taken, we would add; namely, the demon must be still further humanized by sharply limiting his power of statistical averaging and inference when confronted with the particular conditions under which many biological processes occur. These conditions are not amenable to conventional statistical treatment. The aim of this monograph is to show that this limitation introduces a larger indeterminacy and radically changes the character of the permissible inferences.

We will develop specific implications of this last statement. We turn now to other quotations counter to the preceding one, to guide us more directly toward the kernel of the argument of this monograph.

Several physicists and other scientists have pointed to the nondeterministic nature of living processes, even though they have stopped short of drawing immediate implication to the subject matter of biology. (See references to the works of Bohr, Elsasser, Weizsaker, Weiss, Commoner, Glass.) The following excerpt from Heisenberg (1962, p. 155) may be selected from among the writings of various physicists who have explored the significance to biology of the basic ideas of quantum theory. "It may well be that a description of the living organism that could be called complete from the standpoint of the physicist can not be given, since it would require experiments that interfere too strongly with the biological functions."

On other grounds, somewhat the same point of view has been taken by many classical biologists in opposition to deterministic explication (Weiss, 1959; Commoner, 1961; Glass, 1963). Among these biologists is Weiss (1959), from whom a statement may be selected to complement the one above. "Of course, the assumption that biology could ever reach the physical end completely is a delusion, based either on lack of realistic acquaintance with living systems and their true nature or unawareness of the conceptual limitations of physical reductionism." These biologists have employed various arguments to show that the laws governing biological phenomena are more general than those of classical chemistry and physics. Like the physicists, these biologists have insisted on a nonvitalistic interpretation of living processes, but have stopped short of demonstrating which assumptions may not be strictly met in the application of physical postulates to living processes.

This monograph resolves the counterposed views of determinism and vitalism by examination of the degree to which biological phenomena may ultimately receive mechanical explication. The resolution takes an unexpectedly positive form, *a broader application of quantum theory for explication of emergent living processes.*

The new proposition that is stated in this monograph is that the phenomenological conditions under which minimally essential classes of processes occur in living units are such that the uncertainty product of conjugate variables of molecular motion is enlarged at ordinary temperatures and pressures so that molecular motion in these biological processes is mechanistically indeterminate. This molecular indeterminacy engenders new laws for living processes and must be taken into account in a new classification of living beings.

Our thesis, then, is that emergent phenomena of biology have unexpected explication under a broad application of quantum reasoning to molecules, which reflects the phenomenological conditions of many crucial living processes. The following statements serve as a preview of the kinds of considerations and terms in which the new science of biology will be couched if minimally necessary classes of processes in living units can be plausibly shown to occur under phenomenological conditions which do not permit utilization of the central limit theorem of probability with its subsidiary assumption of Gaussian distribution of conjugate variables of molecular motion in an indeterminacy relation.

1. The assumption is justified of supra-mechanical stability of stationary states in atomistic or discrete living units, which we refer to as "bions." (Bions are not equivalent to cells. An organized cell is not a bion, but is composed of bions. Bions are a more fundamental living unit than cells.) (See Chapters 5 and 6.)

2. A new classification of biological organisms is developed from a consideration of basic biological units or bions. (See Chapter 6.)

3. The existence of long-range, specific, additive, electromagnetic forces with exchange of virtual photons is predicted in the development of bonds for crucial biological coascervates and aggregates which bridge the gap between small molecules and large molecular complexes such as bions. The significantly longer range of these additive forces depends on the special conditions of molecules which do not permit assumption of Gaussian distributions of oscillators. Transitional probabilities and exchange energies are increased between oscillators. (See Chapter 7.) Under these conditions, in an unpredictable manner and below a critical temperature, certain emergent collective properties come into existence. One of these properties is the existence of long-range forces, without which the bion or cell could not exist. Examples of similar properties in nonliving material are superfluidity and superconductivity.

4. Changes of stationary states of these living units or bions are predicted as radiation phenomena, analogous to spectral radiation of atoms. (See Chapter 8.) This radiation is predicted to be coherent, originating from bions and difficult to detect except by bions or by very special sensors such as Liesegang rings (see Glossary). It may take a particle or wave form. A biological situation may be discovered in which there occurs emission and absorption of bionic radiation (such as "mitogenetic radiation" [Rahn, 1936]) by living units or bions with change of energy states. This is analogous to a nonliving situation, such as emission and absorption of black-body radiation by glowing atoms with change of energy states. (Some of the difficulty with authentication of mitogenic radiation in the past may have resulted from overlooking the presence of coherence or incoherence in superposition of waves. Laser beams have some analogy here.)

5. Transformations similar to those employed in cybernetics are useful as "quantum" or selection rules for describing the most general forms that discrete solutions of quantum mechanical equations may take for indeterminate processes which comprise bions. (See Chapters 4 and 5.) Although not enough is known to construct these transformations completely, the forms of the classes of transformations are specifiable according to the phenomenological characteristics or conditions under which many essential classes of processes occur in living units. At least two classes of processes are essential for living units or bions: "reproductive" (information conservation) processes and "metabolic" (mass and energy transport) processes. Another class includes "epigenetic" (information conservation) processes, as these processes appear in more complex biological organisms. These various classes of transformations have sub-classes or parameters.

6. The information conservation processes are closed cyclic processes, or radically recurrent in nature. For genetic processes, this is described by a general class of transformations that are single-valued with identical intervals or single-valued with addition of linear intervals. For epigenetic processes, this is described by a general class of transformations which are single-valued with addition of logarithmic or combined logarithmic and linear intervals. The mass and energy transport processes are radically open, asymmetric, or nonequilibrium in nature. These metabolic processes are described by a second general class of transformations that are non-single-valued (see Chapter 5).

7. The significance of the above is that explication of living processes has forced a view in which the fundamental concepts of motion of bodies and of interaction of motion of bodies are synthesized or inductively conjoined in a manner sufficiently general to encompass situations in which time variation and sequence of nonequilibrium states are important, as well as those in which time variation and sequence of steady (essentially static or equilibrium) states may be assumed.

8. A gene may be considered as an element of a particular parameter or subclass of the general class of transformations for genetic information transport (conservation). A hormone may be considered as an element of a particular parameter or subclass of the general class of transformations for metabolic mass and energy transport in a given bion which operates in a molecularly indeterminate way through selection of the working parameters of any of the general classes of transformation of other individual bions or bions in multibionic cells (see Chapters 5, 6, and 9).

9. Biological mutations are elegantly explicable as irreversible changes of bions to new informational states. Viewing bions as atomistic units, mutations become analogous to radioactive transformations of atoms.

10. Transformations for information transfer may be thought of in terms of genetic exclusion principles for bions, as the Pauli exclusion principle is for atoms. The potential for creation of biological niches may be thought of as the operation of an evolutionary pressure or force. A choice in expression of this potential in the form of which ecologic niche will be filled may be thought of as the operation of epigenetic selection rules.

11. In Part II the bion is discussed more extensively in Chapter 6 where viruses are referred to as sub-bionic particles. This terminology is used because viruses do not qualify as bions under the minimal capability requirements of reproduction and metabolism. They require other bions or living cells for metabolism. Most generally, viruses are analogous to subatomic particles of atoms. As Lwoff (1966) first reported, some viruses, such as lysogenic bacteriophages, disappear into the genome of a living cell only to appear unpredictably at later times. Oncogenic viruses are passed in tissue culture without evidence of the virus' actually living in cells. What viruses do when they enter living cells or bions is analogous to what alpha particles do when they are

bombarded into nuclei of atoms. The nuclei of atoms change energy states and eject some similar particle. Rickettsia are of bionic size, have a single plasmic envelope, but require a cellular environment for growth. They may be interpreted as analogous to chemical free radicals which are trapped in a frozen matrix. They seem to be "parasitic" or degenerate bions. Both viruses and rickettsia exhibit infectivity or capability for entrance into cells, which may be very specific.

12. The pervasive Law of Initial Value in physiology and the equally pervasive Arndt-Schulz Law of stimulatory effect of small amounts of poisons in pharmacology seem to coincide in form with the opposite "knees" of the multivalued or discontinuous curve of operation of negative resistance elements. (See Chapter 9.) The significance of this is that the broadened application of quantum theory accounts directly for several of the most important phenomenological laws of metabolism already known in biology. The negative resistance elements of electronics (Mackay, 1958, 1960), such as a relaxation oscillator, which have a multivalued or asymmetric nature, serve as useful analogies with the open processes of metabolism, for instance, membrane irritability or sarcomeric contraction, which have a multivalued or asymmetric nature.

13. A formal decision-making policy for the factor, $\in$, for enlarging the uncertainty product in the uncertainty relation in keeping with non-normal statistical conditions would be satisfactory in order to indicate the level of organization in living processes at which quantal explication *must* be employed. Thereby, a common bridge of explication would be completed among living units or bions with different organization and behavior in various organisms. However, short of this, it remains satisfactory that various atomistic units emerge with an indeterminate basis of operation, depending on the size of a factor, $\in$, which is related to the phenomenological conditions of restricted numbers of particles in small spaces and relatively short times, and of openness of mass and energy interchange with environment. In other words, it is quite useful to recognize that generalized complementarity requires a broader "working" statement concerning quantization of action under the uncertainty relation and appropriate quantification of the correspondence rule if it is to be generally applicable so as to include processes with phenomenological conditions which necessitate assumption of other-

than-Gaussian, as well as Gaussian, distributions of conjugate variables in an indeterminacy relation.

14. The seemingly negative experiments of Nathanshon (1900) and the Chambers (1961) are evidence for the validity of the broader application of quantum theory in biology in much the same way that the seemingly negative experiments of Michelson and Morley (1887) are evidence for the validity of the special theory of relativity (see Chapter 10).

The experiments of Nathanshon do not indicate that there are deterministic molecular mechanisms operating in the cellular nucleus in genetic information transport in reproduction. These experiments demonstrate the constancy of genetic information transport *despite mitotic* or *amitotic* divisions of chromatin macromolecules of the nuclei of *Spirogyra* cells. These experiments are evidence that genetic information transport is quantal in nature. The closed cyclic transformations under which information transport processes are subsumed in a broader application of quantum theory are capable of describing the stability of genetic information transport in *Spirogyra* despite the lack of any classical mechanism.

The experiments of the Chambers are negative with respect to interpretation in terms of random motion of molecules of the role of living cellular protoplasm in mass and energy transport in metabolism. The Chambers demonstrated the open (asymmetric) nature of mass and energy transport in metabolic processes by detecting pH's of 7.6 ± 0.2 in the nucleus and pH's of 6.8 ± 0.2 in the cytoplasm of living echinoderm eggs and other cells. In addition, the absence of Brownian or random motion of enzyme molecules in the nucleus and cytoplasm indicated the asymmetric nature of the metabolic processes. When the echinoderm eggs or other cells were micropunctured, pH fell to about 5.4 in both nucleus and cytoplasm, coincident with the onset of Brownian motion of the cytoplasm and nucleus and death of the egg or cell. These experiments provide evidence that there is a necessary asymmetry and nonequilibrium for maintenance of metabolic integrity in living cells. The multivalued transformations under which the open nature of mass and energy transport processes is subsumed in a broader application of quantum theory are capable of describing the asymmetry or nonequilibrium of metabolic mass and energy transport which is necessary in living echinoderm eggs and other biologic cells.

15. The point of this monograph is not intended to be negative,

designating that classical and statistical mechanics and ordinary quantum mechanics do not apply in biology. The application of these descriptions at appropriate levels is valid, but their sufficiency for many living processes is questioned. (At this point we should also like to call attention to quantum statistical mechanics which biologists have employed in only a few instances so far but which ought to have important applications in biology.)

The thesis of this monograph allows an affirmative answer to the question of whether atoms and molecules exist as such in the "bion" and in biological cells. The general rules for behavior of atoms and molecules can not be extrapolated, however, from behavior of molecules and atoms en mass, i.e. in large numbers. An element of enlarged indeterminacy is introduced which extends beyond electrons and atoms to encompass molecules. Statistical concepts of physics and chemistry that are based on the assumption of the central limit theorem and a sequence of equilibrium or "steady" states do not appear to be warranted for the peculiar phenomenological conditions of many essential biological processes.

16. This broadened application of quantum theory, while sacrificing mechanistic determinism, introduces the possibility for formal explication of many baffling holistic processes of biology, for example, sexual conjugation, nuclear and extranuclear inheritance, and epigenesis. (See Chapters 6 and 9.) Using this broadened application permits hormone function and enzyme action to be approached more meaningfully than is otherwise possible. New insights arise into embryogenesis and many of the important quantum-related disease processes of man, such as carcinogenesis, congenital malformation, aging, essential hypertension, atherogenesis, collagen-vascular hypersensitivity, psychosis, vicarious calcification, etc. A deeper understanding and positive knowledge of the nature of the world of small actions is gained, including many of the crucial processes of biology. This deeper understanding constitutes the basis of rules limiting descriptions of the world of small actions, but also constitutes the basis of rules implicitly applying nondualistically for all physical actions.

These general points are expanded in the various sections of the text. Because of the nature of the subject matter, biologists may find the following references (listed below) useful supplementary reading: Bohr, Reichenbach, Elsasser, Born (Chaps. 1, 4, 5, and 8), Brillouin (Chap.

21), Polanyi (pp. 139–42), Heisenberg (p. 155), Goodwin, Ashby, Wiener, Lindsay and Morgenau, Khinchin, Weizsaker, von Neumann (p. 2070), Burnside (p. 80), Burgers, Patte, and Agar. For physical scientists, the following writings are suggested: Chambers, Commoner, Glass (p. 223), Weiss (p. 449), Williams, Thompson, Heilbrunn, and Lamanna and Mallette.

References

Agar, W. E. 1951. 2d ed. *A Contribution to the Theory of Living Systems*. Melbourne: Melbourne University Press.

Ashby, W. R. 1963. *In Introduction to Cybernetics*. New York: Wiley.

Bohr, N. 1934. *Atomic Theory and the Description of Nature*. London: Cambridge University Press.

———. 1958. *Atomic Physics and Human Knowledge*. New York: Wiley.

Born, M. 1962. *Atomic Physics*. 7th ed. New York: Hafner.

Brillouin, L. 1962. *Science and Information Theory*. New York: Academic Press.

Burgers, J. M. 1965. *Experience and Conceptual Activity*. Cambridge: M.I.T. Press.

Burnside, W. 1959. *Theory of Probability*. New York: Dover.

Chambers, R., and Chambers, E. 1961. *Explorations into the Nature of the Living Cell*. Cambridge: Harvard University Press.

Commoner, B. 1961. *Science* **133**: 1745.

Elsasser, W. M. 1958. *The Physical Foundations of Biology*. New York: Pergamon.

———. 1961. *J. Theoret. Biol.* **1**: 27.

———. 1962. *J. Theoret. Biol.* **3**: 164.

———. 1964. *J. Theoret. Biol.* **7**: 53.

———. 1966. *Atom and Organism*. Princeton: Princeton University Press.

Glass. B. 1963. In *Philosophy of Science—The Delaware Seminar*, Vol. 1, ed. B. Baumrin. New York: Interscience.

Goodwin, B. C. 1963. *Temporal Organization in Cells*. New York: Academic Press.

Heilbrunn, L. V. 1956. *The Dynamics of Living Protoplasm*. New York: Academic Press.

Heisenberg, W. 1962. *Physics and Philosophy*. New York: Harper.

Khinchin, A. L. 1949. *Mathematical Foundations of Statistical Mechanics*. New York: Dover.

Lamanna, C., and Mallette, M. F. 1965. *Basic Bacteriology*. 3d. ed. Baltimore: Williams and Wilkins.

Lindsay, R. B., and Margenau, H. 1957. *Foundations of Physics*. New York: Dover.

Lwoff, A. 1966. *Science* **152**: 1216.

Mackay, R. S. 1958. *Am. J. Phys.* **26**: 60.

———, 1960. *IRE Trans. Med. Elect. ME–***7**: 94.

Morowitz, H. J. 1960. In *Medical Physics*, ed. O. Glasser. Chicago: Year Book Publishers.

Nathansohn, A. 1900. *Jahr., Wiss. Bot.* **35**: 1.

Neumann, J. V. 1956. In *The World of Mathematics*, ed. J. R. Newman, New York: Simon and Schuster.

Pattee, H. H. 1967. "Quantum Mechanics, Heredity, and the Origin of Life." Publication CTS–B–67–1. Coral Gables, Fla.: University of Miami Center for Theoretical Studies.
Polanyi, M. 1958. *Personal Knowledge*. Chicago: University of Chicago Press.
Rahn, O. 1936. *Protoplasma—Monographien* vol. 9. Berlin: Verlag von Gebruder Borntraeger.
Reichenbach, H. 1944. *Philosophic Foundations of Quantum Mechanics*. Berkeley: University of California Press.
Thompson, D. W. 1963. *On Growth and Form*. London: Cambridge U. Press.
Weiner, N. 1948. *Cybernetics*. New York: Wiley.
Weiss, P. 1959. In *Biophysical Science*, ed. J. L. Oncley, New York: Wiley.
Michelson, A. A., and Morley, E. W. 1887. *Am. J. Sc.* **34**: 333.
Weizsaker, C. F. v. 1955. *Naturwissenschaften* **42**: 531.
Williams, R. J. 1963. *Biochemical Individuality*. New York: Wiley.

PART 1

DEVELOPMENT OF QUANTUM HYPOTHESIS FOR BIOLOGY

CHAPTER 1

MOLECULAR INDETERMINACY

Molecular indeterminacy in many essential living processes

This monograph proposes a previously unsuspected application of quantum theory in biology. Under the complementarity principle, phenomenological conditions which do not permit assumption of Gaussian distribution of conjugate variables of molecular motion in many essential living processes rule out determinacy of motion of the molecules constituting these processes. Conversely, statistical mechanics is not appropriate for explication of many essential biological processes for precisely the same reason. Statistical mechanics concerns behavior of atoms and molecules en masse, a situation in which the phenomenological conditions of a process permit assumption of the central limit theorem of probability theory with Gaussian distributions as the basis of the statistical description of the process. Specifically, under the complementarity principle, the uncertainty relation may be employed to establish mechanistic indeterminacy for molecules in view of the phenomenological conditions under which the processes occur. Mechanistically indeterminate description must be complementary in nature, and the principle of anomaly requires that such description must have a quantum feature. Quantum theory is open-ended and can accommodate a mechanics for quantization of molecular motion. This permits a nonvitalistic, nondeterministic, unexpectedly quantum explication for large molecular action, as the basis of many necessary processes in living units or organisms. A correspondence rule, under generalized complementarity, places a boundary upon this broader quantum description precisely at limiting conditions where statistical mechanics is appropriate. This qualifies quantum description of molecular motion in these emergent processes as nondualistic with respect to classical physical theory.

The first step in formulation of a fundamental theory for explication of molecular motion in processes of living organisms is to appreciate the nature of physical theory. Physical theory is concerned most fundamentally with motion of bodies and interaction of motion of bodies. This is to say that description of physical processes is ultimately concerned with or may be reduced to the description of mechanical action or mass and energy transport in space and time in correspondence with the laws of motion. Several different modes of description of mechanical processes are appropriate, depending upon whether measurements do or do not significantly interfere with the putative action of a process. In one mode, in which measurement does *not* significantly interfere with the putative action of a process, Newtonian mechanics sets down criteria for complete, that is predictive or postdictive, description. These criteria imply simultaneous measureability of conjugate variables of motion. This is equivalent to stating that processes which are describable under classical mechanics are mechanistically determinate. Under appropriate phenomenological conditions, statistical limit theorems of probability theory may be utilized as the basis for a statistical mechanical description of certain processes. Statistical mechanics, employing the central limit theorem, is a complete description of these mechanistically determinate processes. When measurement *does* interfere significantly with the putative action of a process, the conjugate variables are complementary in that the width of the distribution of measurements of one conjugate variable is reciprocally dependent upon the width of the distribution of measurements of the other conjugate variable. These processes are mechanistically *indeterminate* and necessitate quantum description, because the principle of anomaly prevents the unambiguous construction of interphenomena during intervals of measurement. This is a larger mode of description which includes classical mechanics and statistical mechanics in the limit.

This mode of complementary description of mechanistically indeterminate processes has a necessarily probabilistic interpretation and corresponds with the mode of complete description of mechanistically determinate processes at some limit under a correspondence rule which requires nondualism with classical physical theory.

The second step in formulation of a fundamental physical theory for biology is to establish the minimum number of most general, necessary classes of living processes. Heuristically to merit the descriptive term, living, it is essential that, at the least, two classes of processes are associated in living units. These classes are here entitled "metabolic" and "reproductive." These classes must be included because they are exhibited by all living units except viruses (about which some question exists concerning status as living units) and possibly ricksettsiae (see Chapter 5). "Metabolic" pertains to energy and mass transport in a broad sense, not just to events of molecular catalysis in intermediary metabolism. "Reproductive" pertains to information conservation in a broad sense, not just to events of molecular replication in chromosomes. The principle characteristics of these two classes of biological processes may be stated briefly. Metabolic processes are radically *open* or asymmetric regarding mass and energy transport in space and time to and from the environment of a living unit. Reproductive processes are radically recurrent or *cyclic* regarding limitation of numbers of large molecules of a living unit in time and space.

The third step in the formulation of a fundamental physical theory for biology is to establish the appropriate mode of description of mechanical action of the above minimum number of classes of most general, essential processes in living units.

The outstanding feature of the development of biological theory in the past several centuries has been the increasing recognition of the microparticulate[1] foundation of biological processes. The critical point to be stressed in this regard is the proscription of conditions of essential classes of processes in basic living units in the form of small number (low concentration) of large molecules in small spaces for short periods concurrent with open transport of mass and energy to and from the environment at ordinary temperatures and pressures. Radically large is the number of energy states which are accessible in phase space through which the point can pass that represents the metabolic state of a living cell during necessarily open or asymmetric transport of mass and energy to and from environment. Concurrently, there is a restriction in number

[1] Electrons, atoms, ions, molecules, and some aggregates of these.

of energy states which are occupied, that is, through which the point actually passes that represents the reproductive state of a basic living unit during necessarily cyclic restriction of number of molecules in time and space.

The central limit theorem of probability theory is useful for summation of a sequence of an extremely large number of random, mutually independent variables in the form of a normal or Gaussian distribution of probabilities. However, variables may not be permitted the description of "independent" in a sequence which is intended to describe the trajectory of the probabilistic state of a system through accessible energy states of phase space when the proscription of conditions under which that system generates the sequence requires asymmetry, i.e. radically large number of energy states that are accessible for occupancy during generation of the sequence. Variables may not be described as "statistically large in number" in a sequence which is intended to describe the trajectory of the probabilistic state of a system through accessible energy states of phase space when the proscription of conditions under which that system generates the sequence requires cyclicity, i.e. restriction in the number of energy states of the system that are occupied during generation of the sequence. Both types of conditions lead to insufficient numbers of occupied energy states. The meaning of this proscription is that the central limit theorem and other statistical limit theorems of probability theory may not be employed as appropriate asymptotic formulas for assumption, in the uncertainty relation, of reciprocal Gaussian distributions of conjugate variables of action of large molecules in metabolic and reproductive processes. This establishes indeterminacy of the action of large molecules in these classes of minimally essential processes in living units.

Descriptions of processes which are indeterminate must be complementary, and energy and momenta of mechanistically indeterminate processes must be quantized. Furthermore, quantum explication under the above phenomenological conditions is nondualistic, with respect to classical mechanism, and is not boundless, since at a limit of appropriately large numbers of occupied energy states the explication corresponds with statistical mechanical description.

The new feature in living processes is unexpected indeterminacy for large molecules, which is explicitly engendered by the phenomenological conditions of these processes. The conditions under which many biological processes occur are such that there is an insufficient number of occupied energy states for use of statistical limit theorems. *The point can not be overstressed that this indeterminacy does involve the trajectory of large molecules, whereas indeterminacy that involves phenomenological conditions under which statistically adequate ensembles of particles may be assumed is usually restricted to the trajectory of electrons and atoms.* The latter is true in that Planck's constant or quantum of action is the minimal uncertainty product for microparticulates in the uncertainty relation for most nonliving processes. This is the case in most nonliving processes, since the phenomenological conditions are fulfilled or may be fulfilled experimentally for a sufficiently large number of occupied energy states, and the central limit theorem with reciprocal Gaussian distributions of conjugate variables of microparticulate motion may be appropriately assumed.

A mechanics for molecular processes in living units is seen to be qualified in this manner by an increased indeterminancy which yields an advantage, namely, explication of *emergent* processes in living units under a broader application of quantum theory.

It is worth repeating that any currently available analytical procedures, such as the statistical limit theorems of probability theory, which are utilized to bridge the connection between molecules of living cells and observable classes of processes of living cells must be looked upon with appropriate restraint. Although description of such connections under statistical mechanics (which depends on such extremely large numbers of microparticulates that the nature of the microparticulates and the chemical and physical laws of interaction between the microparticulates may be disregarded) seems satisfactory for given mechanistically determinate models of processes of living cells, the burden of proof that the chemical and physical conditions of a model actually obtain in a processes of a living cell remains unrelentingly with the model builder. The connections cannot be assumed because of the assumptions of the model. Furthermore, the consistency of a number of models, when a separate model is posed for each of the myriad of individual living processes, is questioned unless the phenomenological conditions which permit statistical assumptions for all the individual models are simultaneously applicable in the living unit or cell. The crux of the situation is that the phenomenological conditions of small numbers of large molecules in small spaces and short times and open transport of mass and energy with environment at ordinary temperatures and pressures cast serious doubts on whether statistical mechnical explication can ever be demonstrated as applicable for many processes which are necessary for living cells. To summarize: the phenomenological conditions for two essential classes of processes of a living unit, metabolic and reproductive, preclude the use of the central limit theorem of probability theory (*a*) in the assumption of Gaussian distributions in any uncertainty relation which is applied to the processes and (*b*) in statistical mechanical descriptions of these processes. *A* establishes these processes as indeterminate, necessitating complementary descriptions with quantization of energy in stable action states, of which the most stable states are living atomistic units (bions; see Chapter 5). *B* implies a limit under a correspondence rule at which description of the processes is nondualistic with statistical mechanics.

Of course, there are alternative forms of explication.

With the ruling out of the central limit theorem, a first tendency is to try other limit theorems of probability theory in a modification of statistical mechanics, but these also depend on phenomenological conditions which permit assumption of a large number of occupied energy states.

Any attempt to redefine the phase space of the organism has the inherent likelihood of being "non-physically realizable" (Rosen, 1962). Such a recodification, even though analytic and affording predictions would have to be considered as a dualistic theory unless it is specifically shown to be equivalent to a physically realizable transformation. This difficulty arises because there exists a large number of ways to map the transformations embodied by an organism, and it is by no means clear how to restrict them arbitrarily. An analogous situation existed several centuries ago when it was thought that there was a biological distinction between "organic" and "inorganic" chemistry.

Elsasser (1966) stresses a view of biology which leads to complete individuality, leaving out all chemical and physical laws to catalogue living functions. Elsasser states that even the totality of all members of a particular species constitute an immensely small number of the possible configurations of their constituent molecules so that use of statistical inference is not permissible in bridging the gap between the microscopic states and the macroscopic behavior of an organism. The comment which may be made about his important, pioneering view is that there seems to be no correspondence rule to afford a connection with classical physics.

To be considered also is the branch of science known as general dynamics which is an attempt to generalize the laws which apply in all mechanical systems. This branch of science has the virtue that it pertains to but is not restricted to phenomenological situations where large numbers of occupied energy states are involved. One of the most important problems of general dynamics concerns the dependence of the character of the motion of any mechanical system on initial data. Khinchin has pointed out that general dynamics is more precisely concerned with the determination of the characteristics which in one or another sense "almost do not depend" on these inital data. Such quantities for a large proportion of trajectories assume values very near to a certain constant number. Such stability considerations may come to be useful, if they can be shown to include quantized systems of a small number of degrees of freedom, as found in living units (Rubin and Sitgreaves, 1954; Gontcharoff, 1944).

Our view is based upon a plausibility argument for indeterminacy in the essential processes in a living unit, which is discussed in Chapter 3. Previously, no one has exploited indeterminacy which extends to motion of large molecular microparticulates in living units (bions) and in relations between living units; that is, interpretation of indeterminacy has been limited to motion of electrons in atoms and in relations between atoms. Perhaps the main reason has been that the conditions of most nonliving processes have been appropriate for utilization of the powerful analytical tools (limit theorems) of statistical inference.

Before going on, several pointed disclaimers must be included to avoid confusion. The implied refutation of statistical mechanics as a descriptive mode for certain physical and chemical processes in biology does not in any way imply quantization of these processes unless at the same time the reasons for negation of the statistical mechanical description are precisely those that give rise to indeterminacy.

At this point in the exposition nothing has been stated about the statistical interpretation of the operators employed in the quantum mechanics. Even though further expansion of the capabilities of the operators is inferred, probabilistic interpretation of the operators is not implied. At no time in this discourse will the operators be connected through a probabilistic type of argument to the physical situations to which they refer. In Chapter 4, the various forms of the classes of transformations which are adopted as quantum rules (for the proper forms that the solutions of the quantum mechanical equations do take) will correspond to the phenomenological conditions of the classes of processes to which the classes of transformations apply. These disclaimers are so important that Chapter 2 is inserted in order to keep the background in physics quite clear.

References

Elsasser, W. M. 1966. *Atom and Organism*. Princeton: Princeton University Press.

Gontcharoff, V. 1944. *Bull. Acad. Sci. U.R.S.S., Serie Mathematique* **8**: 1.

Khinchin, A. I. 1949. *Mathematical Foundations of Statistical Mechanics*. New York: Dover.

Rosen, R. 1962. *Bull. Math. Biophys*. **24**: 375.

Rubin, H., and Sitgreaves, R. 1954. *Tech. Rept. No. 19 A*, Appl. Math. and Stats. Lab. Stanford, Calif.: Stanford University.

CHAPTER 2

CONDITIONS FOR LARGER INDETERMINACY

Larger indeterminacy in processes which have phenomenological conditions that do not allow assumption of normal distributions of conjugate variables.

At the outset, emphasis must be placed upon the fact that quantum theory is open-ended. Methods of application of quantum theory by means of a mechanics are not implicitly specified, as they are in a closed system, like Newtonian dynamics, for example. The point to be made is that the quantum mode of description has very broad applicability which encompasses many essential living processes in an unexpected way.

A good beginning for an examination of the applicability of quantum theory is a consideration of generalized complementarity, which includes (*a*) the principle of indeterminacy or uncertanity (Heisenberg, 1930); (*b*) the principle of anomaly (Reichenbach, 1944); and (*c*) the Bohr (1961) correspondence rule (see Heisenberg, 1930, Chapter 5.)

RELATIONSHIP BETWEEN INDETERMINACY PRINCIPLE AND THE QUANTUM DESCRIPTION WHICH EMBODIES IT

Bohr and Heisenberg advocated a restrictive interpretation of physical processes in which the only descriptive statements that are considered meaningful are those about simultaneously measurable conjugate variables of action. Conjugate action variables of interphenomena which occur between phenomena can in no way be measured simultaneously in the case of physical processes which have putative actions approximately equal to the minimal disturbance engendered by a measurement.

The interphenomena are mechanistically indeterminate and can be described only in terms of complementary variables. "Complementary" here implies that, if a dynamical variable of action is precisely known, its conjugate kinetic variable is reciprocally imprecisely known, and vice versa, and that the product of their suitably defined latitude of uncertainties is equal to or greater than a minimal constant or universal quantum of action (h). This is a statement of Heisenberg's uncertainty or indeterminacy principle, which is a fundamental physical law that holds for all possible physical situations. Its focal importance in physics is apparent, in view of the fact that this same relation may be derived from the axioms of quantum reasoning, or may be demonstrated in conceptual experimental situations involving ideal measurements, or may be predicted qualitatively from the general principle of complementarity without stipulation of the magnitude of indeterminacy.

The world of interphenomena can be introduced only by an interpolation based on definitions. Reichenbach has pointed out in the principle of anomaly that such a supplementation of the world of phenomena can *not* be constructed free from anomalies.

> The causal anomalies cannot be removed because they are inherent in the nature of the physical world. The *principle of indeterminacy* formulates only one part of this nature; it states that it is impossible to verify certain statements about interphenomena. To this is added, by the system of quantum theory, another principle we have called the *principle of anomaly*. It states that no definition of interphenomena can be given which satisfies the requirement of a normal causality; it therefore maintains the impossibility of a normal supplementation of the world of phenomena by interpolation.
>
> The limitations of scientific interpretation of the world of quanta, expressed in these two principles, must not be considered as limitations of the power of the human intellect. It is not human ignorance, nor lack of knowledge, which leads to the conditions imposed upon descriptions of the physical world expressed in the laws of quantum mechanics. It is positive knowledge, deep insight into the nature of the atomic world, which constitutes the basis of this strange network of rules, formulated as rules limiting descriptions, but expressing im-

> plicitly rules holding for all physical occurrences. Beneath the disguise of a theory of physical knowledge we discern the outlines of a physical world different from what centuries of scientific research had dreamed it to be, but nevertheless demanding recognition as the world of reality (Reichenbach, 1965, p. 44).

This fundamental law of the physical world, the principle of anomaly, is derivable from the basic principles of quantum theory, as is the fundamental law of indeterminacy.

The correspondence rule of Bohr is quantified under quantum formalism. The rule is an expression of the fact that as the totalized action of a physical process becomes sufficiently large, a limit is reached at which simultaneous measurement of conjugate variables can be made with sufficient accuracy for the trajectory to be described by classical mechanics.

Delving into the epistemological problem of knowledge and observation in physics, Bohr enunciated the principle of complementarity as a general statement in qualitative terms of the ultimate limits which can be probed into the detailed structure of matter. This was his attempt to resolve the many dilemmas posed previously in the field where actions approach the size of the quantum of action.

Bohr's principle of complementarity (Bohr, 1961, pp. 83–93) states that our knowledge of a physical system or process is always complementary to the measurement of its state. Bohr said that the word "experiment" refers to a situation in which what has been done and learned can be told to others. He stressed that the account of the experimental arrangement and the results of the observations must be expressed in unambiguous language with suitable application of the terminology of classical physics in which space-time (kinematical) concepts and conservation (dynamical) laws of momentum and energy must be utilized simultaneously to make a complete, predictive, or postdictive description of the course of a physical system.

An experiment or measurement to a certain degree deprives the knowledge given by a previous measurement of its significance for predicting the future course of a physical system. A subtle meaning thus appears in the principle of complementarity, namely, that a limit is set upon the extent of knowledge obtainable by measurements and also upon the meaning or interpretation which may be attributed to these measurements. For classical physical systems, space-time concepts and spatial continuity may be assumed, permitting a causal chain of explication. Theoretically, assumption of infinite divisibility of space-time is satisfactory in large or classical physical systems because the graininess or limit of knowable divisability of the physical system which appears with actual or real measurements is insignificantly small in comparison with a large physical system. The principle of complementarity, though still pertaining as a generalization, holds no constriction upon a classical physical system. A limit of knowable divisability of physical systems begins to appear when the space-time divisions of a physical system approach the limit of the division of physical scales and clocks as measuring instruments. Here the important words are "knowable" and "measuring." Both knowledge and measurement, as stated above, require expression in terms of space-time concepts *and* dynamical conservation laws simultaneously.

The uncertainty principle expresses this explicitly in pointing out that certainty of measurement of momentum or energy of a physical system cannot be known by an observer, if certainty of measurement of space or time (respectively) is known. In other words, with physical systems a point of divisibility is reached where interaction of a physical system with a measuring instrument constitutes an inseparable whole. Change of experimental or measuring arrangements changes the state of such an atomistic system in indivisible, incompletely predictable, quantum jumps.

Bohr's principle of complementarity was promulgated under the restrictive interpretation of physical processes in mechanical terms as outlined above and is a metalanguage for interpretation of the epistemological relation between description of a process and measurement of action of the process. If applied in an epistemological analysis, the principle of complementarity *necessitates a feature of uncertainty or indeterminacy* when putative action of a process is approximately equal to the product of unavoidable and expected uncertainties of simultaneous measurement of conjugate mechanical variables. Estimate of the magnitude of the uncertainty product of a process depends explicitly on the extent of *unavoidable* and *expected* uncertainties of simultaneous measurement of conjugate variables, (*a*) which extent of unavoidable uncertainties is related to interaction between the process and measuring device and (*b*) for which extent of expected uncertainties no meaning or latitude can be defined except on the basis of averaging and probabilistic inference.

At this point, and *within a latitude of expected uncertainties of measurement of conjugate variables, suitably defined in magnitude* under assumption of reciprocal Gaussian distributions of conjugate variables, the unambiguous minimal expression of a physical system equals a universal quantum of action, $h/4\pi$. It is at this brink that the constraint of the principle of complementarity on extent of knowledge from physical measurements and meaning of physical measurements is expressed (*a*) in the manner of complementary descriptions when experimental conditions are changed so either kinematical or dynamical variables may be more precisely known, (*b*) with loss of classical determinism, and (*c*) in necessarily quantal terms. This resolves in a single frame the paradox of many different predictive descriptions of the same physical system.

It bears repeating that the uncertainty principle is a point of departure which allows a new insight into physical description of many living processes, since the uncertainty relation may be derived from (*a*) quantum mechanical formalism, or (*b*) demonstrated by *gedanke* considerations of many different experimental systems with ideal measurements, or (*c*) predicted qualitatively by the principle of complementarity. The point of departure is with respect to the "suitably defined latitude of uncertainties of measurement" of conjugate variables of action of a physical process.

PHENOMENOLOGICAL CONSIDERATIONS CONCERNING THE STATISTICAL ASSUMPTIONS IN AN INDETERMINACY RELATION

This focuses upon phenomenological considerations which undergird the statistical assumptions in the uncertainty or indeterminacy relation itself. The latitude of expected uncertainties of simultaneous measurement of conjugate variables can be suitably and minimally defined in magnitude only if reciprocal Gaussian distributions[1] of conjugate vari-

[1] The virtue of Gaussian or normal approximation to binomial distribution of probabilities is that this distribution is an exponential, continuous, mathematically symmetrical (bell-shaped) distribution, which represents to a good approximation the distribution of the arithmetic mean of a large number of independent, random variables, so as to define the minimal possible latitude or dispersion of expected uncertainties about a single, most probable or expectation value, the arithmetic mean.

ables may be appropriately assumed in making an estimation of the minimal magnitude of the uncertainty product.

The question must be raised whether the phenomenological conditions of many processes in living units warrant the isomorphic assumption of Gaussian distributions of conjugate variables.

Most nonliving physical processes[2] qualify for appproriate isomorphism in that phenomenological conditions of these processes fulfill, or can be made experimentally to fulfill, the criteria of (*a*) describability in terms of conjugate variables of mechanical action, (*b*) sufficiency of number of microparticulates, and (*c*) sufficiency of closedness of transport of mass and energy in space and time with respect to environment (with sufficiency in *b* and *c* relating to summability under the central limit theorem). The isomorphism with Gaussian distributions is obvious when the subsidiary assumptions of Gaussian distributions are appreciated to be (*a*) unequivocal assignability of events to logically disjoint classes, i.e. equiprobability of events, (*b*) sufficiency of number of events, and (*c*) sufficiency of independence of events (with sufficiency relating to summability under the central limit theorem).

This may be applied, for example, to the indeterminacy expression: $d(q) \cdot d(p) = h/4\pi$, where the product of uncertainty of position $d(q)$ (defined as a Gaussian distribution of all positions) and uncertainty of momentum $d(p)$ (defined similarly as a reciprocal Gaussian distribution of all momenta) is equal to h divided by 4π (h is 6.6×10^{-27} erg sec.).[3] If the phenomenological conditions do not fulfill the criteria for isomorphic assumption of reciprocal Gaussian distributions of conjugate variables, the uncertainty relation for such a process no longer holds as an equality, with $h/4\pi$ as its product. *For expression as an equality*, the uncertainty product must be multiplied by a factor $\in$ that will depend on some decision making policy which is based upon a consideration of the phenomenological conditions of the process. $\in$ will have the value 1 only when the physical conditions are such that Gaussian distributions may be appropriately assumed.

[2] (Certain processes are excluded, for example, some meteorological and nonreversible processes such as nuclear, explosive, high energy plasma, and other nonequilibrium reactions.)

[3] For a clear derivation of the uncertainty relation from quantum postulates the reader may consult Dicke and Wittke, p. 129–132, *Introduction to Quantum Mechanics*, Addison-Wesley Publishing Co., Inc., Reading, Massachusetts, 1950, or Born, p. 401, *Atomic Physics*, Hafner Publishing Co., New York, Reprinted 1962, 1963.

PHYSICAL CONDITIONS WHICH PRECLUDE THE ASSUMPTION OF GAUSSIAN DISTRIBUTIONS OF CONJUGATE VARIABLES IN AN INDETERMINACY RELATION AND ENFORCE A LARGER INDETERMINACY PRODUCT

Metabolic processes are radically open in transport of mass and energy to and from environment at ordinary temperatures and pressures. Living processes of reproduction involve small numbers of large molecules in small spaces and relatively short times. Physical conditions of both of these essential classes of processes in living units explicitly restrict the sufficiency of the number of occupied energy states and thereby preclude assumption of Gaussian distributions of conjugate variables of motion of molecules in these classes of processes. The plausibility argument of Chapter 3 provides justification for these statements.

REQUIREMENT BY PRINCIPLE OF ANOMALY OF QUANTAL DESCRIPTION OF PROCESSES WHICH HAVE AN INDETERMINATE BASIS

Processes with physical conditions which force values of $\in$ which are greater than 1 in the indeterminacy relation and have putative action approximately equal to $\in \cdot h/4\pi$ have indeterminate bases of operation, and processes which have indeterminate bases must have a quantal feature in their description, *in view of the principle of anomaly*, a point well developed by Reichenbach (1944, Part 1).

WHAT MICROPARTICULATE DIMENSIONS DOES THE ENLARGED QUANTAL DESCRIPTION ENCOMPASS?

By *gedanke* considerations, the indeterminacy product may be enlarged by virtue of three kinds of changes of phenomenological conditions of physical processes: (*a*) the number of molecules can be smaller, (*b*) the size of the molecules can be larger, and (*c*) the motion of the molecules can be slower. In minimally essential living processes, smaller numbers and larger sizes of molecules are found to be present. In a physical situation *c* is accomplished by a great decrease in temperature. At the temperature range of living processes, the motion of neither molecules nor electrons is varied significantly enough to force quantization. Remaining within the temperature range of living processes, the dimensions of microparticulates which are next encompassed when heuristically another level of action is quantized are those of molecules.

Not knowing the size of $\in$ for living processes, we can not state directly that the larger indeterminacy encompasses motion of molecules. This forces us, for the present, to examine carefully and state the phenomenological conditions of living processes (in Chapter 3) in order to establish plausibly the hypothesis that the larger indeterminacy does encompass molecules.

REQUIREMENT BY CORRESPONDENCE RULE OF NONDUALISM OF ENLARGED QUANTAL DESCRIPTION WITH CLASSICAL PHYSICAL THEORY

The question arises of where the broader applicability of quantum theory is supplanted by the known applicability of statistical mechanics. This transition may occur when the events comprising the process become sufficiently numerous or approximate an equilibrium condition. In other words, if the process is amenable to utilization of the central limit theorem, its description is supplanted by statistical mechanics or ordinary application of quantum theory (for processes occurring with phenomenological conditions, so the minimal indeterminacy product may be assumed). The deciding factor between the latter two is whether the measurement of conjugate variables of microparticulate motion interfere significantly with the putative action of the process. This fulfills the requirement under a correspondence rule for nondualism of the broader application of quantal theory with present physical theory.

Here is seen the elegance of generalized complementarity with the combined power of the principle of indeterminacy, the principle of anomaly, and the correspondence rule, when all are generalized. Generalized complementarity is the metalanguage that makes possible a discussion of just such a nonconventionally statistical situation wherein the indeterminacy product is enlarged under conditions which do not allow assumption of normal distributions of conjugate variables. For this monograph the point is crucial, because many essential living processes in living units occur with just such conditions.

The above considerations suggest a classification of physical processes which is presented now for the purpose of clarity in discussion of the new class of physical processes which is proposed in this monograph.

CLASSIFICATION OF PHYSICAL PROCESSES[4]

Three classes of phenomenological conditions of physical processes can exist in biology or elsewhere, and can be used in classifying physical processes under mechanical descriptions.

The process is amenable to description in classical mechanical terms (and hence, is mechanistically determinate), if the action being considered in a process is much larger than $\in h/4\pi$, with $\in$ either equal to or greater than 1, i.e. the distribution of conjugate variables may be assumed to be either Gaussian or non-Gaussian. As an example, even though only one event (the extremely non-Gaussian condition) is being described, the action possessed by (the center of gravity of) a moving truck is so much greater than the product of uncertainties engendered by measurement of the positions and momenta of its atoms that gross estimate by classical mechanics would enable a bystander to know its position and velocity and avoid being run over. Statistical mechanics, which has an equally mechanistically determinate basis, applies when all actions of a process are larger than $\in h/4\ \pi$and the distributions of conjugate variables may be assumed to be Gaussian under the appropriate utilization of the central limit theorem.

The second class of phenomenological conditions is similar to the first except that the action is very small and of the same order of magnitude as the product of uncertainties. A careful qualification is necessary of the distributions of conjugate variables of the process. If they may be appropriately assumed to be Gaussian, $\in$ is 1 in the uncertainty equation, stated as an equality with $\in h/4\pi$. Here the processes are amenable to descriptions in quantum mechanical terms. Such conditions only occur at atomic spatial dimensions and time intervals because of the extremely small magnitude of $\in h/4\pi$. An example is the action possessed by an electron in the orbit of an atom, or an atom in a molecule.

The third class of phenomenological conditions is that which is of most interest in this monograph, namely, the class such that Gaussian distributions of conjugate variables of action may *not* be appropriately assumed and the uncertainty product $\in h/4\pi$ is enlarged. ($\in$ is greater than 1.) $\in h/4\pi$ can enlarge until it becomes equal or comparable to some reasonable classical action being considered (an action large enough, that, if considered under the central limit theorem with as-

[4] Further discussion of this topic is to be found in Brillouin. 1962. P. 230.

sumption of Gaussian distributions, it would have fitted under statistical mechanics). Examples would be the action of a molecule in a macromolecular aggregate or a marcromolecular aggregate in a bion. A new bridge emerges for making quantal explications of physical systems at different levels of molecular action.

A plausibility argument must now be presented in Chapter III to substantiate the crucial point that Gaussian distributions of conjugate variables may not be assumed for the motion of molecules in many essential processes in living units, because of the nature of the phenomenological conditions which establishes a larger indeterminacy in these processes.

References

Bohr, N. 1961. *Atomic Physics and Human Knowledge.* New York: Science Editions.

Brillouin, L. 1962. *Science and Information Theory.* 2d ed. New York: Academic Press.

Heisenberg, W. 1930. *The Physical Principles of the Quantum Theory.* New York: Dover.

Reichenbach, H. 1944. *Philosophic Foundations of Quantum Mechanics.* Berkeley: University of California Press.

CHAPTER 3

LARGER INDETERMINACY IN BIOLOGY

Hypothesis that phenomenological conditions of many essential (metabolic and reproductive) processes in living units preclude the assumption of normal distributions of conjugate variables of molecular motion

In the last chapter, it was shown that certain distinct conditions must be met before the central limit theorem can be utilized to justify the assumption of a normal distribution for the conjugate variables of motion in physical processes. There must be a sufficient number of particles, or alternatively the time interval possible for observation must be long. At the same time the process must be sufficiently closed. The aim of this chapter will be to demonstrate that generally the biological cell metabolizes and carries out the transfer of genetic information in a manner so as to preclude the assumption of these conditions and, in turn, to preclude a normal distribution of the conjugate variables of molecular motion. Hence, by virtue of the uncertainty relation, many of these processes are indeterminate. This type of argument has been pursued, as was pointed out in Chapter 2, since at present we do not know the magnitude of $\in$ and cannot use it in a direct investigation of each process.

THE UNIVERSAL LIMITATION OF NUMBER OF MICROPARTICLES IN BIOLOGY

Every living unit, including unicellular organisms, coenocytic plants and multicellular organisms, reproduces within the boundaries of a small volume, e.g., the cell, at some point in its life history. This rephrasing of the cell doctrine simply emphasizes that genetic information is transferred by processes occuring with restrictions on absolute numbers of molecules. The size limitation can be more strongly made when the concept of the bion is introduced in Chapter 5, since larger cells are composed of these units which retain replicating capability (bions).

Each cell contains in its structure many different kinds of large molecules. This perforce limits the absolute number of each kind. This conclusion is reinforced by data based on vitamin synthesis and enzyme induction (McIlwain, 1946). McIlwain reports that riboflavin in *Aerobacter aerogenes* is produced at an average rate of 1.4–2.0 molecules per bacterium per second. Biotin in the same organisms is produced at 0.08–0.34 molecules/cell/sec and p-aminabenzoic acid in *Escherichia coli* is manufactured at about 3.3 molecules/cell/sec. *In vitro* systems of enzymes are able to produce these vitamin molecules at rates of 100 molecules per enzyme molecule per second or higher. The possibility is that only one such molecule exists per cell (Hogness, 1959, pp. 256–58). The type of conclusion about small numbers of enzyme molecules in a cell is consistent with enzyme induction studies. Beta galactosidase is present in *E. coli* which are grown on glucose. If grown on lactose, 1000 times the amount of enzyme is required and inductively synthesized. This amount now represents 1 per cent of the protein of the cell. The dry weight of protein in *E. coli* is 4.7×10^{-13} gm. so that before adaptation the enzyme comprised about 5×10^{-18} gm. Converting the molecular weight, 400,000, to actual weight, one enzyme molecule weighs roughly 7×10^{-19} gm. By this estimation there are about 9 enzyme molecules per cell.

As Setlow and Pollard (1962, p. 46) have pointed out, "This leads us to the remarkable idea that for some factors . . . the whole process must take place *in the vicinity of one molecule*. Where this molecule is in the cell, how its substrate can reach it, and the considerable consequences of the formation of a second molecule are excellent subjects for thought. It is also clear that any process that occurs at a single molecule is not one to which the statistics of large numbers can be applied."

Large holistic processes in living units are frequently influenced by only a few molecules or ions. Substances thought of as "triggers" and catalysts such as antigens, activators of physiological processes, hormonal moderators, vitamins, trace minerals, and certain toxins all extend their action in low absolute concentration. Their relative concentration per target cell is, then, extremely low. It has been estimated, for example, that: (*a*) Only one molecule of the ragweed antigen "E" is necessary per receptor cell for immunological action (King, Norman, and Connell, 1964). (*b*) One molecule of Hageman factor is required for initiation of blood clotting (Wald, 1965). (*c*) Around 2000 molecules of botulinum toxin are fatal to a mouse (Lamanna, 1959). The molecule is a specific neurotoxin, affecting cholinergic motor end-plates, so that this number aligns with some number of critical units paralyzed. Granting that a critical muscle like the diaphragm may be the target, it is quite unlikely that paralysis of one end-plate is lethal, and not likely that paralysis of 100 end-plates is lethal. Even though paralysis of some number of end-plates may be lethal, as this number is assumed to be larger and larger, fewer and fewer molecules could possibly be involved in the paralyzing process at any one end-plate. (*d*) The zinc concentration in human leucocytes has been determined to be 3.2×10^{-10} micrograms per million cells (Hoch and Vallee, 1952). This concentration represents about 0.2 atoms of zinc per cell. Should all the zinc be concentrated in one of the least common subfractions, for example, the basophils which comprise about one half of one per cent of the leucocyte population, the value would be increased to about 40 atoms per cell. This number does not seem to be the result of chance inclusion, since it bears a constant ratio to the amount of protein in normal cells and falls to 10 per cent of this value in persons with leukemia.

Lacking knowledge of turnover rates and localization patterns, it is difficult to establish quantitatively the number of hormone molecules per effector cell. Angiotensin, serotonin, aldosterone, catecholamines, thyroid, and especially the peptide hormones, either in absolute concentration in plasma or in incremental response, exert effects in which there could be only a relatively few hormone molecules per cell in a target organ or tissue. By way of contrast, it may be pointed out that the chemist working at his present equipment limits—microscopic volumes, 10^{-6} microliters, and moderate concentrations, ca. 10^{-3} M, or very low concentration, 10^{-10} M, and moderate volumes, ca. 10 microliters—is still concerned with the statistical behavior of about 10^{10} molecules.

TIME LIMITATION OF MOLECULAR MOTION

Every living cell has a definite upper time limit on its existence. The life span of the cell then constitutes the maximum time interval in which to accumulate a time average in a presumptive measurement of molecular motion relating to genetic transfer of information. This time may be relatively short as in some bacteria (ca. 20 minutes) or very long as in the case of cells that persist for the lifetime of a multicellular organism. Such cells, though they replace components within themselves, typically lose the ability to reproduce another cell.

The lifetime of the cell, while constituting the maximum interval, probably does not at all represent time available for the completion of a molecular event leading to the preservation of a genetic character. This would be true only if all the molecular events were to occur in a parallel manner. Any set of reactions that occur serially would correspondingly further limit the time interval associated with the molecular species concerned. For example, Puck and Marcus (1956) have shown that DNA synthesis occurs during about 5 per cent of the life cycles of typical mammalian cells.

Every cell seems to cycle its molecular components. The composition of the cell is at any instant an exhibition of incessant intracellular processes. Radioactive tracer studies, beginning with Schoenheimer (1942; see also Eagle, 1959) have shown this to be true of all types of molecules including the deoxyribonucleic acids. The length of time that a particular molecule can be engaged in events concerned with reproduction is thereby limited. Yet, the cell's integrity is dynamically maintained. With respect to metabolism, Eagle (1959) studied protein turn-over by culturing mammalian cell lines of both normal and malignant variety. "When cultures are overlaid with a medium lacking one or more essential amino acids, there is of course no net synthesis of protein. Isotopically labeled amino acids are nevertheless incorporated from such media into the cell protein at the surprisingly high rate of approximately 0.7 to 1.0 per cent per hour. The reverse process (that is the loss of amino acids from prelabeled cell protein) proceeds at the same rate, provided that the corresponding amino acid is present in the medium. This incorporation is largely an intracellular process and is only in small part due to either cell degeneration or protein secretion into the medium, followed by degradation to the amino acid level and resynthesis. It is noteworthy that metabolic inhibitors such as azide, dinitrophenol, chloramphenicol, 6-mercaptopurine, and 5-fluorodeoxyuridine, at concentrations which totally inhibit the net synthesis of protein, have no significant effect on the rate of protein turnover."

THE RADICALLY OPEN NATURE OF METABOLISM

Chemical Interactions

Metabolic interactions taking place within the cell can be divided into three major groups: *substrate pathways*, *catalytic environment*, and *cell milieu*. This grouping is made in order to manage the discussion economically. It does not imply a functional significance to the grouping. These groups are represented by blocks in Figure 1. Figure 1 suggests the reciprocal interactions that may occur between elements of each group.[1] In addition to these groups of natural interactions there is another group, *viz*, *enzyme separation techniques*. A large part of our knowledge concerning the cell interior and its biochemistry comes from this interaction. The block labeled "substrate pathways" refers to the concentration and pathways of chemical interconversions involving relatively small molecules. The cell milieu refers to the physico-chemical properties of the solution (cytoplasm, nucleoplasm, etc.) in which the interactions take place. Catalytic environment is a term introduced by Grisolia (1964) to refer to the behavior of an enzyme during its catalytic action. This term includes the plastic-elastic nature of the active site of the enzyme (Koshland, 1959), the contribution of the secondary and tertiary structure of the protein, states of aggregation of various subunits, and the possible energy transfer mechanisms over the protein and its surface (Klotz et al., 1958, and Lumry, 1959). By setting its action apart in this manner there is less tendency to regard the enzyme as a passive partner in a reaction sequence.

Figure 2 may be viewed as an enlargement of a portion of Figure 1, picturing in somewhat more detail the reciprocal interactions occurring in two of the systems. The diagram is not to be taken as literally true for any one enzyme system, but rather as a collection of known effects for various enzymes. In turn, neither diagram is inclusive, in that large segments of interaction such as *compartmentalization* and *enzyme induction* are not pictured.

[1] It is not merely coincidental that this diagram is similar to the one given by Ashby (1963) to represent the coupling of four or more machine-like elements with feedbacks. Ashby states that the behavior of such a system cannot be predicted from the complete knowledge of the dynamic characteristics of each of the elements. This is to say that the coupled system exhibits emergent behavior. This is precisely the point to be developed here. Indeed, it would seem that this emergent behavior of a multiply connected system could be made the basis of a parallel but equivalent argument to the one presented here.

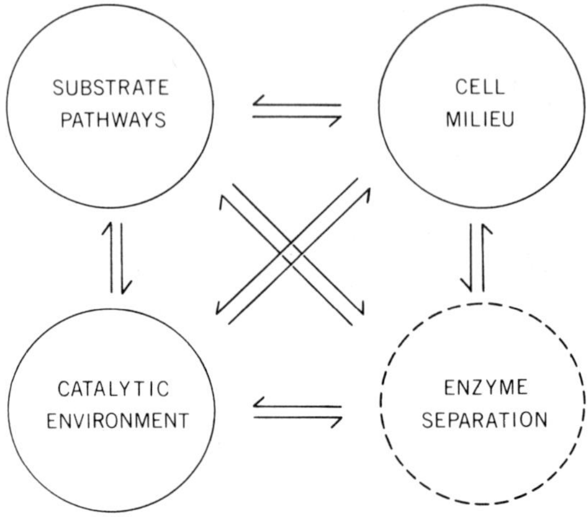

Figure 1. Some major groups of metabolic interactions.

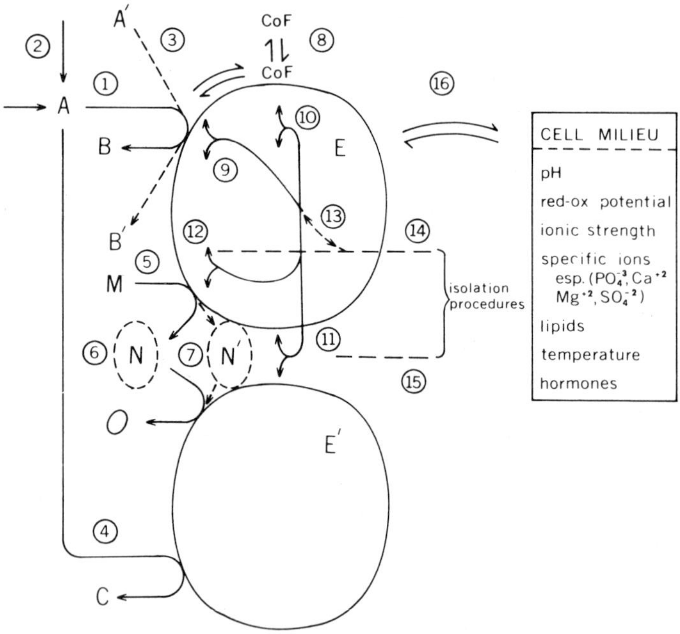

Figure 2. Some possible couplings between substrate pathways and catalytic environment in the neighborhood of two enzymes in situ. External arrows *indicate substrate pathways involving the active sites of the enzymes.* Internal arrows *represent the contribution of the plastic nature of the protein molecule to catalysis.* Dashed horizontal lines *represent effects of isolation procedures.* (*Individual arrows representing reciprocical influences from each of the items in the rectangle are omitted for clarity*).

The large number of different metabolic pathways existing within the cell does not of itself engender the *radical openness* necessary to preclude statistical treatment. It can be shown (Wilkie, 1960) that a *finite sequence* of chemical reaction, if provided with a constant input of reactants of arbitrary concentration, will within a finite time settle down to the steady state. This steady state is characterized by a zero rate of change of concentration of reaction intermediates, by a minimum rate of entropy production, and by a theoretical description in terms of successive approximations to equilibrium. If, on the other hand, two such reaction sequences are coupled, for example by a common intermediate, the time interval required to reach the steady state increases enormously or becomes infinite. (In the last instance, the coupled system displays a type of behavior known as relaxation oscillation (Mackay, 1960) in which the concentration of the common intermediate alternately swings between high and low values.) Using the language of statistical me-

---

Key for Figure 2

Substrate Pathways

① Simple catalysis exhibited by enzyme, E, isolated by procedure ⑭.
② Substrate A production by several pathways.
③ Competition *or* promotion of ① by *in situ* alternative substrate at the same active site, e.g. cyclic oxidation-reduction of a group at the action site.
④ Participation of A in alternative pathway, A→C.
⑤ Reaction M→N exhibiting multifunctional nature of enzyme E if separation procedure ⑬ does not intervene.
⑥ Participation of the two enzymes in a simple consecutive reaction in which intermediate N is kept low by a rapid N→O.
⑦ Adsorption of intermediate N′ whose reaction rate is radically affected by separation ⑭.
⑧ Cofactor(s) directly affecting rate of ①.

"Internal" or structural stability factors of a plastic enzyme

⑨ Activation or inactivation of enzyme by presence of substrate A.
⑩ Activation or inactivation of enzyme by cofactor.
⑪ Activation or inactivation of enzyme by adsorption to or complexing with other proteins, carbohydrates, lipids, etc.
⑫ Polyvalency and specificity of enzyme influenced by ⑨, ⑩, or ⑪.
⑬ Reciprocal influence of isolation and activation.
⑭ Isolation procedure which ruptures relatively strong bonds or causes extensive dissociation of protein subunits, e.g. disulfide bond reduction or translocation.
⑮ Isolation procedure disrupting relatively weak bonds such as adsorption to a common matrix or dispersion interaction.
⑯ Cell environmental influences capable of affecting substrate reaction rate, substrate specificity, enzyme activation-inactivation, and the isolation procedures.

Arrows might be drawn separately to each item in the rectangle and the site of interaction on the enzyme and also between each item labelled the cell milieu.

chanics, the number of possible states of the system has increased upon coupling to the extent that the states which are actually occupied form a very small fraction of the total possible states.

In general, it can now be stated that coupling between the various pathways of metabolism tends to increase the openness of the metabolic system as a whole and, in fact, makes it radically open whenever the coupling is as extensive as within the cell.

Evidence for the large degree of coupling that occurs between the classical "static" metabolic pathways is of two types. The first may be considered to be indirect, deriving, as it were, from an extrapolation of the known behavior of the protein enzymes *in vitro* to the conditions known or thought to exist within the cell. Evidence of the second type is somewhat more direct, consisting of behavior of the cell that could only occur by virtue of multiple coupling. These two types of evidence will be examined in this order.

The list accompanying Figure 2 is intended to delineate in a systematic manner some of the interconnections of the first type. Some of the molecules known to be involved in multiple pathways are those of the energy-yielding substrates, e.g. glucose, fatty acids, and some amino acids, cofactors and energy exchange intermediates, such as the nucleotide phosphates, and common metabolic intermediates, such as acetyl coenzyme A and the folic acid intermediates of the one-carbon pool. This list could be multiplied many-fold by a few minutes perusal of a set of metabolic "maps," but serves to introduce the nature of the interconnections in the substrate pathways.

For some enzymes there must occur a considerable change in structure and energy during interaction with a substrate. This forms the basis for Koshland's (1959) "induced fit" hypothesis for enzyme action. Grisolia (1964), beginning with the observation that interaction with substrate could render an enzyme either more *or* less sensitive to inactivation, has extended a general theory of the "plastic-elastic" nature of the enzyme to a large class of important related considerations. He gives instances in which enzyme specificity, multivalency states of aggregation, and reaction rates can all be markedly influenced by the degree of interaction with substrates, cofactors, and inhibitors. These changes in the catalysts of the cell affect large changes in the dynamics of the coupled substrate pathways and further extend the possible dynamic couplings in metabolism. Direct evidence that the postulated couplings exist in the intact cell comes from a variety of sources. The autotrophic and photosynthetic organisms that use a single carbon source obviously distribute the carbon to all synthetic, storage, structural, and energy supply systems. Radiolabled carbon in substrates shows the same type of rapid distribution in heterotrophic cells, though not exactly to the same extent as in autotrophic cells. The action of hormones upon the cells of multicellular organisms is another example of the intimately coupled nature of metabolism. One hormone can show multiple effects upon a target cell. Davidson (1965) shows, for example, that ACTH, thyroxin, and insulin directly influence gene activation and at the same time show action(s) clearly independent of gene activation. Conversely, the same effect may be demonstrated by several hormones. For example, the completely different chemical molecules, epinephrine and glucagon, exert almost identical effects on liver phosphorylase (Stetten, 1959).

Anesthetic agents are of a wide variety, ranging from temperature and simple ions to complex molecules (Heilbrunn, 1943). They are able to elicit essentially two responses from a comparably wide range of cells—raising the irritability at low concentrations or lowering the irritability at higher concentrations. This multiplicity of response from one agent, or its converse, may be taken as *prima facie* evidence of coupling of the type inferred from knowledge of the possible interactions.

Transport Processes

As a result of the *continual* metabolic exchange of mass and energy, molecules in the interior of the cells are subjected to (and at the same time lend their share to maintaining) dynamic potential gradients. The word "dynamic" is deliberately chosen in order to include those forces arising in transport phenomena. Irreversible conduction of mass (diffusion or flow), energy (thermal conduction), momentum (viscosity), or charge (electric current) gives rise to a reversible potential gradient for the other three "fluxes" (Eisenschitz, 1958). When charge is involved, the phenomenological effects are collectively known as Hall effects (Barnothy, 1960, Vol. 3, pp. 61–4). Otherwise each specific effect is named, such as the Soret effect or Dufour effect. Each of these dynamic potentials may suffer a discontinuity at an interface. For thermal gradients, the interface may act as an insulator or enhanced conductor;

mass may suffer positive or negative adsorption; shear gradients exist for viscosity; and zeta potentials modify charge transport. These interactions take on especial importance in the biological cell since the surface to volume ratio is extremely high due to the existence of numerous membranes and particles of colloidal dimensions.

These effects, arising only in a dynamic fashion, can now be seen to contribute to the radically open nature of metabolism. The coupling among the various fluxes themselves and in turn the coupling involving the substrate pathways, catalytic environment, and cell milieu serve to increase the number of accessible microstates of the system.

Phenomenological manifestations of coupling among the various transport processes are the dynamic asymmetry of ion gradients and microstructural stability of the cell. Two series of experiments detailed by the Chambers illustrate this point directly (Chambers and Chambers, 1961, pp. 5–34 and 141–74).

The pH of the cytoplasm of various cells, determined by microinjection of sulfonphthalein indicators, is found to be constant, i.e. pH 6.8, regardless of the acidity of the external medium, as long as the cell is viable. Interruption of respiration by metabolic inhibitors and coagulation of the protoplasm by injection of Ca^{++} or by cytolysis uniformly resulted in a lowering of the pH to 5.4 ("the acid of injury"). This response also was independent of the external acidity within wide limits. While the exact interpretation of even the most carefully made acidity measurement in concentrated colloidal media is open to question (Feldman, 1956), the constancy of the measurement and its consistent fall upon cessation of metabolism is evidence of the dynamic maintenance of an ionic gradient. In a related series of experiments, Brownian movement of the various cytoplasmic inclusions was not observed in the cell until the cytoplasm was subjected to cytolysis by divalent cations or micropuncture. Upon inhibition of respiration and cell death the dynamic structuring of the cytoplasm ceases, and the particles are subjected to the random thermal bombardment characteristic of Brownian motion.

The following quotation from Bohr (1961) states well the case for asymmetry or nonequilibrium of energy handling in living processes.

> Notwithstanding the essential importance of the atomistic features, it is typical of biological research, however, that we can never control the external conditions to which any separate atom is subject to the

> extent possible in the fundamental experiments of atomic physics. In fact, we cannot even tell which particular atoms really belong to a living organism, since any vital finding is accompanied by an exchange of material through which atoms are constantly taken up and expelled from the organism which constitutes the living being. Indeed, this exchange of matter extends to all parts of a living organism to a degree which prevents a sharp distinction on an atomic scale between those features of its mechanism which can be unambiguously accounted for on usual mechanics and those for which a regard for the quantum of action is decisive.

A longer list of special points could be introduced to bolster the plausibility argument, but the preceding generalizations are sufficient for the argument, in that they cannot be dismissed as explicable under mechanistically determinate theorems of chemistry and physics which must be applied in a mechanics that does not destroy the living processes.

Cells Without Metabolism

Metabolism is a continuous characteristic of living processes in all cells. This is evinced by the gaseous respiration of "resting" cells, by the depletion of endogenous substrate in starved cells, and by disappearence of phosphate intermediates from metabolically poisoned cells. There are cells, however, that are able to enter a state in which there is virtually no metabolism. Such cells, as for example, the endospores of bacteria, dormant embryos of seeds, and peripheral tissues of hibernating animals, show none of the characteristics usually associated with living processes. Growth, irritability, and movement are suspended or seem to be suspended in order to enter a state capable of structural stability under adverse environments. When these conditions for the apparent exception to the ubiquity of metabolism are duly noted, they are seen to cause no problem in the argument that has been pursued.

In summary, the most general aspects of cellular metabolism point toward its necessary description as radically open. The conditions of continual, multiply-coupled exchange of mass and energy and the resulting asymmetry between cell and environment combine to increase the number of possible microstates of the metabolic system. The number of states occupied over any time interval remains small and thereby insufficient for the application of the central limit theorem.

OTHER FORMS OF PLAUSIBILITY ARGUMENTS

The plausibility argument against utilization of the central limit theorem and Gaussian distributions of conjugate variables of molecular motion in many living processes is supported indirectly by a series of other arguments. These other arguments are based on the observations that (*a*) some phenomena of biology appear to be quantized; (*b*) other phenomena are most easily explained in quantal terms; and (*c*) still other phenomena can be shown to be indeterminate (quantal). Item *c* may be examined through a consideration of necessary precision of measurement of conjugate variables to insure a mechanistically determinate basis of operation, together with a demonstration by a calculation with an indeterminacy equation of the inability to achieve this necessary precision by physical means.

With respect to *a*, Thompson (1942, Vol. 1, pp. 170–1) points out that growth of *Spirogyra* (after Hofmeister) and of the crocus plant (after Bose) appears to be quantal.

With respect to *b*, phenomena of biology which are easily explained in quantal terms include the occurrence of the M space (see page 00, below) in bacterial growth (Lamanna and Mallette, 1965, pp. 370–1), the nonminimal character of viscosity of bacteria at the isoelectric point (ibid., p. 63), the Arndt-Schulz Law concerning the tendency of poisons in very low concentrations to stimulate rather than depress biological processes (ibid., pp. 897–901), the Gram staining reaction in bacteria (ibid., pp. 162–76), the amitotic division of *Spirogyra* (Nathanshon, 1900), the coding ambiguity in polypeptide synthesis with bile salts or alcohol without change of DNA or transfer conditions (Davis *et al.*, 1965), mitogenetic radiation (Rahn, 1933), long-range forces in biological complexes (Hoffman, 1957, pp. 288–306), endogenous or exogenous (*Zeitgeber*) rhythms in biology (Sollberger, 1962; and Brown, 1959), stratification of amebae by centrifugation with survival (Heilbrunn, 1943, pp. 65–67), virus particle entry, disappearance, and reappearance in cells (Lwoff, 1966), and the physiological Law of Initial Value (Wilder, 1962).

With respect to *c*, processes of biology, such as genetic information transport can be shown to be indeterminate (hence quantal) if a demonstration of necessary accuracy and amount of information transport is made, and application of an indeterminacy relation is utilized to show that this could not be accomplished by any physical means of information transport which has a mechanistically determinate basis (see later in this chapter). This demonstration and application in an uncertainty relation may even be made assuming Gaussian distributions of conjugate variables and still indicate indeterminacy. A similar demonstration and application can be made for an important metabolic process. An example is given concerning material conduction of excitation to the interior of a skeletal muscle fiber (see later in this chapter).

These additional forms of arguments are taken up in Chapters 6–10. It may be productive at this point for the reader to skim quickly through the gist of these sections in order to appreciate the formidable character of these additional plausibility arguments. The whole monograph must be reviewed, once perused, as a cogent argument itself for the non-Gaussian quantal theme in living processes. The predictions, the unification of many seemingly diverse explications, and the nondualistic nature of the explications are further considerations. Finally, the reasonable establishment of the results of a broadened application of quantum theory in many living processes also strongly supports the plausibility argument of non-Gaussian distribution.

LIVING PROCESSES IN WHICH MOLECULAR INDETERMINACY CAN BE DEMONSTRATED

Up to now our argument has been concerned with demonstrating that the most general conditions under which biological processes occur tend strongly to limit the assumption of a normal distribution of conjugate variables of the motion of molecules comprising these processes. As a consequence, the enlargement of the uncertainty product (with the impossibility of complete measurement and description) forces such processes to admit only of a quantal description. The very small magnitude of the uncertainty product of normally distributed variables, i.e. approximately Planck's constant, has previously resulted in the tacit limitation of processes in which quantum effects predominate to those of atomic dimensions. In other words, a line of demarcation at atomic dimensions has been created between processes necessitating quantal description and those amenable to classical description.

In consideration of the temperature and energy of inputs consistent with life, statistical mechanics has been usually assumed valid, commencing with molecular motions and higher. The assumed placement of the boundary between classical and quantum description pertaining to cell processes is justified when almost infinite time or space is allowed. However, as shown above, very few cellular processes occur under conditions in which a stationary or quasi-stationary state may be assumed. It is not surprising to find material transport processes in biology in which the rigorous specification of time or other intervals lead to indeterminacies of molecular motion in spite of the assumption of normally distributed conjugate variables. Two examples are presented below.

> Several comments should be made about these sample calculations. One, the plausibility argument of this chapter does not depend on these calculations. The calculations do strengthen the plausibility argument, but the previous points are sufficient to establish the plausibility that phenomenological conditions in many essential living processes do not permit assumption of Gaussian distributions in an uncertainty relation for the processes. Two, these particular calculations are presented not only because they demonstrate the necessity for considering complementarity in several classes of processes which are essential in living organisms, but because these calculations make the demonstration *even while assuming Gaussian distributions* in the uncertainty relations, i.e. without applying the main theme of this monograph. Three, the reader may wish for more sample calculations which employ pairs of complementary variables. Biological experiments have not often been conducted from the view of complementarity. There is a paucity of data which may be utilized in uncertainty calculations, regardless of the skill of the biologist attempting the calculations. Four, had previous experimentation in biological processes taken complementarity into consideration, it is probable that a conclusion concerning quantization of molecular motion would have occurred earlier.

One of these sample calculations is relatively straightforward and deals with the material conduction of excitation to the interior of the skeletal muscle fiber. Specification of the momentum and distance limits within which the process must occur can be made so as to show the excitation process to be indeterminate.

The other example, involving hereditary information transfer in a small bacterial cell, is introduced for two reasons. The first reason is that it serves heuristically to further the argument of Linshitz (1953, pp.

14–15) and especially Polonski (1964, pp. 481–502) that information within the cell cannot be separated from the molecular motions. The second reason is that it shows, in a selected case, that this information with its requisite accuracy of transmission cannot be completely specified molecularly.

It must be emphasized that the choice of examples depended on their adventitiously falling within the compass of the minimum uncertainty relation, that is, without $\in$ being larger than 1. They are only intended to bolster the previous argument of this chapter by demonstrating that there are, indeed, indeterminate processes at the molecular level within the cell.

In this connection, Brillouin (1962, pp. 230–40) gives a set of criteria for a non-perturbing observation. In the first instance, the accuracy of the experiment must not be too high. This is to say that the product of the deviations of coordinates, $\Delta_{\text{observ}}\, q$, and momentum, $\Delta_{\text{observ}}\, p$, which are allowed in order to adequately test the hypothesis, must be far from the limits of the uncertainty relations i.e. $\Delta_{\text{observ}}\, p \cdot \Delta_{\text{observ}}\, q \gg h$.

> This use of the triple inequality sign, while undoubtedly correct, does not provide an unquestionable standard when used alone in cases where the product, $\Delta_{\text{observ}}\, p \cdot \Delta_{\text{observ}}\, q$, is some 2–100 times larger than h. We have accordingly omitted examples of this type, and used only several in which the product was actually less than h.

In the second instance, the system must contain many particles and a large number of quanta. This is precisely the condition which cellular processes do not satisfy and is, in a somewhat modified form, the basic thesis of this monograph.

INDETERMINACY IN THE CURRENTLY ACCEPTED MODEL FOR EXCITATION—CONTRACTION COUPLING IN SKELETAL MUSCLE

The biochemistry, physiology, and morphology of skeletal muscle has been intensively studied. The limits set by the confluence of these investigations permit the following estimation of indeterminacy. Since the work function of muscle is singular and outwardly directed, there is exhibited a remarkable longitudinal orientation extending from the form of the muscle to the active protein arrays themselves. The term sarcomere designates the cylindrical contractile units that occur between the Z lines in a linear array along a myofibril (Figure 3). The dimensions of the sarcomere do not vary appreciably in those fibers that have been investigated among widely divergent animal phyla (Perry, 1956). Their diameter ranges from 0.5 to 2.0 microns and as Perry states, "... No doubt to some extent these figures reflect differences in technic rather than real differences in diameter." The length of the sarcomere, which is not of direct concern here, is somewhat greater than the diameter—about 2.0 to 3.0 microns in the resting fiber. The sarcomere is surrounded in a cage-like fashion by elements of the endoplasmic reticulum (Porter and Palade, 1957). This recticulum, as its name implies, is a plexus of tubular elements which has been shown to have multiple connections with the sarcolemma, the membrane which invests the muscle cell (Rayns, Simpson, and Bertand, 1967).[2]

The muscle cell, more commonly called a fiber, is much more variable in size—typically 10 to 100 microns in diameter and up to several centimeters in length. It is made up of a few to several thousand myofibrils. To complete the picture to the macroscopic level, the muscle body consists of a large number of fibers, along with blood vessels, nerves, and connective tissue.

Skeletal muscle of vertebrates typically exhibits a rapid rate of shortening, or a build-up of tension if it is not allowed to shorten. The interval from stimulus to maximum tension is, to be sure, highly variable between muscles but is characteristic for a given species, muscle type, and temperature. Values for contraction time range at least from 800 to 7.5 milliseconds (Spector, 1956, p. 295). Rat diaphragm, a moderately fast muscle, has a contraction time of 16 milliseconds at normal body temperature. This will be taken as the limiting interval for the following calculation since this muscle has been shown to have an endoplasmic reticulum (Fatt, 1953, and Martin, 1966).

Skeletal muscle normally contracts *in situ* under the impress of a signal or impulse from a motor nerve. Figure 4 depicts the classical and modern views of the route taken from the nerve to the interior of the cell. The

[2] No structural or functional differentiation is made here between the transverse tubular system and the sarcoplasmic reticulum, though they are thought by some to be distinct. Recognition of the lack of structural continuity would simply imply the interposition of another junction in the excitation pathway. A further functional distinction probably implies differentiae which are impossible to observe as is shown here. Accordingly, both systems have been included in the term endoplasmic reticulum, which refers to the internal membranous tubular elements common to most cells.

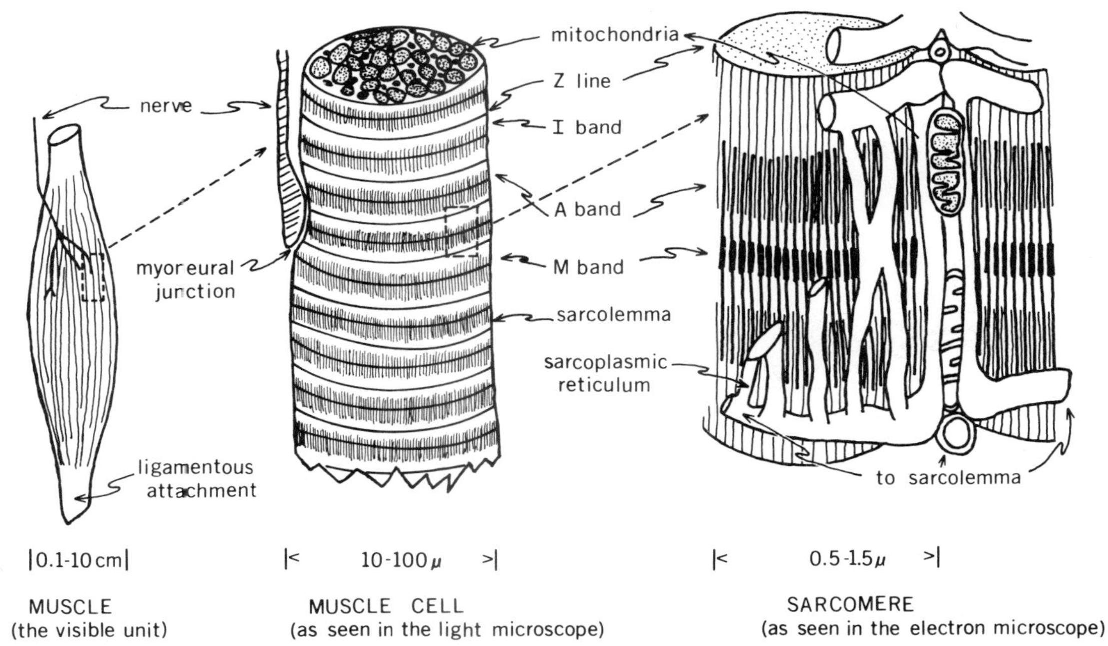

Figure 3. Skeletal muscle at successive stages of magnification (semi-diagrammatic). Since the diagram is intended for orientation, many features such as blood vessels, grouping of muscle cells into bundles, and nuclei have been omitted.

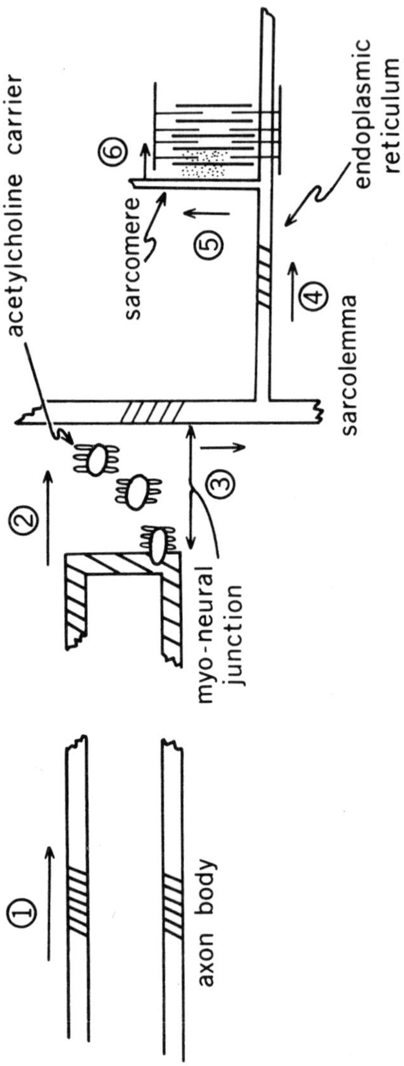

Figure 4. Signal pathway between motor neuron and skeletal muscle. The hatched areas depict the depolarization wave traveling in the direction of the arrows.

propagated impulse, which for our purpose may be simply taken to represent a region of altered ionic permeability, travels down the nerve axon ① until it reaches the myoneural junction, a region of close juxtaposition of the nerve terminal and a specialized region on the surface of the muscle cell. Here ② the impulse is carried across the membrane discontinuity by a transmitter substance, acetylcholine, in association with large protein molecules (Heilbrunn, 1956). A propagated impulse is again set up in the sarcolemma ③ which travels down and around the muscle fiber. The impulse is carried inward by the endoplasmic reticulum ④ and ⑤ and is distributed in turn to each of the contractile units. Certain ions are released by the depolarization of the endoplasmic reticulum in the vicinity of the sarcomere ⑥. Among these ions calcium is thought to be of focal importance. These ions diffuse under either a concentration or electrical gradient or both to the interior of the sarcomere where they initiate the chemical and physical events of contraction proper.

Events ④ through ⑥, while constituting a reasonable hypothesis for the inward travel of excitation, are inferential and are the result of a picture composed from several lines of experimental evidence. Among these may be mentioned: the action of calcium on protoplasm in general and contractile processes in particular (Heilbrunn, 1956), the results of microinjection of calcium ions into muscle fibers (Chambers and Chambers, 1961, pp. 119–40), inward spread of subminimal electrical stimulus only at the Z-line where the endoplasmic reticulum contacts the sarcolemma (Huxley and Taylor, 1958), inferences from the osmotic, permeability, and electrical behavior of the endoplasmic reticulum (Girardier *et. al.*, 1963), and the uptake of Ca^{++} by isolated endoplasmic reticulum in the presence of adenosine triphosphate.

As stated above, this model for the final events in excitation is inferential. Several steps should be followed in utilizing an uncertainty calculation to ascertain whether a proposed mechanistically determinate model such as this is compatible in describing a particular process. First, a specification must be set upon one of the complementary variables. The process itself must be consulted as to maximum uncertainty (imprecision of observation or measurement) of one of its complementary variables. (In the nerve-muscle transmission example above, biological data were considered concerning time in setting the maximum uncertainty in momentum. This value is not arbitrary).

Second, the minimum uncertainty of the other or complementary variable is then given by the uncertainty relation. This is to say the second variable must always have an uncertainty greater than a derived value. A third step is necessary if the applicability of a mechanististic model in a description of a particular process is to be evaluated by an uncertainty calculation. The maximum imprecision in the observed value of a complementary variable in *that* particular process must be considered in order to decide whether the model can be compatible with the minimum uncertainty in the calculated value given by the uncertainty relation.

Recipe for evaluating a mechanistic model with uncertainy relation.

A. From model decide the maximum uncertainty (or imprecision) of variable for such a model to function (e.g. $\Delta_{\text{observed}}\, q$).

B. Next, from the uncertainty relation, obtain the minimum uncertainty of the variable (e.g. $\Delta_{\text{calculated}}\, p$) which is conjugate to the above variable.

C. Compare this calculated minimum uncertainty value ($\Delta_{\text{calculated}}\, p$) with that of the maximum uncertainty (imprecision) of the conjugate variable which is allowed by the model ($\Delta_{\text{observed}}\, p$).

As it stands, the model provides a definite maximum time interval, the contraction time[3] during which a particle must gain at least certain incremental velocity and travel a definite distance to the interior of the myofibril. Using the cited figures, the incremental gain in velocity of a calcium ion acting as a carrier of excitation must be at least

$$\frac{0.5 \times 10^{-4} \text{ cm.}}{0.016 \text{ sec.}} = 3.1 \times 10^{-3} \text{ cm./sec.}$$

This corresponds to a momentum for the calcium ion of

$$\frac{40}{6.02 \times 10^{23}} \times 3.1 \times 10^{-3} = 20.7 \times 10^{-26} \text{ gm. cm./sec.}$$

Therefore, in order to determine whether or not Ca^{++} (or any other smaller ion) does indeed carry the excitation impulse inward, an ideal measuring arrangement—minimum interference, correctable scattering, sufficient number of ions to give a Gaussian distribution of velocities,

[3] The actual interval under consideration is shorter than this by possibly an order of magnitude. Events ② through ⑤ are included in the contraction time, as are the resultant chemical and mechanical events of contraction. Since there is overlap of these events in the fiber, the conservative nature of the uncertainty estimation is maintained by assigning to ⑥ most of the time.

etc.—must be available to determine the momentum to within this value. In other words, the uncertainty in the incremental increase of momentum of an ion during the contraction interval must not exceed 2.07×10^{-25} gm.cm/sec. There is, however, coupled to this, an obligatory uncertainty in position, Δq, by the indeterminacy principle of

$$\Delta q \cong \frac{h}{\Delta p} \cong \frac{6.6 \times 10^{-27}}{2.1 \times 10^{-25}} \cong 3 \times 10^{-2} \text{ cm. or 300 micra.}$$

This is certainly far beyond the desired range of localization of the ion within the sarcomere.

> A similar estimate in the uncertainty in temporal localization may be made independently using Sonnenblick's (1962) data for cat papillary muscle. This calculation involves a somewhat unsure estimate of the ratio of the energy expended in excitation to that of contraction (Botts, 1957), and consequently will only be outlined here. Using an average volume of the sarcomere of $1\mu^3$, an action product of $10^4\, h$ can be estimated for each sarcomere during a contraction. If the value of ratio of excitation to contraction energy is as low as that quoted by Botts—10^{-3}—an uncertainty of 0.1 sec is engendered in events concerned with excitation.

It may be argued that, thus far, the indeterminacy of a mechanism has been shown which was inferential to begin with. This would serve to demonstrate the inadequacy of the model as it is presently formulated. On the other hand, an assessment of the relative contribution of this model to any other coupling mechanism of roughly comparable energy cannot be made; since indeterminacy also precludes the proving false of a plausible mechanism. It would seem that only in the case that a demonstrably determinate mechanism contributed a complete accounting of the excitation energy could the indeterminacy of the model investigated here be ignored.

INABILITY TO ACHIEVE GENETIC INFORMATION TRANSPORT IN *ESCHERICHIA COLI* BY ANY MEANS HAVING A DETERMINATE BASIS

Upon cell division, each of the resulting cells must contain the same minimal amount of information as did the original cell in the initial part of its life cycle. That this must be so can be deduced from the fact that automata, of themselves, cannot produce novel information nor can they regenerate information completely lost in noise (Elsasser, 1958, p. 125). The stability of a species over many generations indicates that cells do accomplish this transfer of certain minimal information (the genome) necessary to specify structure and function. This is especially evident in unicellular organisms dividing asexually and in the somatic cells of a multicellular organism. Conjugation serves to complicate this description only slightly: the union of two cells can at most double the information, resulting in the required amount for transmission by the zygote. Alternatively, conjugation can serve to make the information doubly redundant. The capacity for inducing parthenogenic reproduction in many species makes the latter alternative by far the more likely.

Mutations represent the occasional adventitious change of information by the organism. External mutagenic agents such as radiation are capable of increasing the mutation rate, but the majority of mutations must be considered spontaneous (Lamanna and Mallette, 1965, pp. 572–96). Spontaneous mutations, resulting from accidental rearrangements of internal molecular configuration, represent the lower limit of accuracy of information transfer from mother to daughter cell. The structural and functional complexity of even the simplest bacterium intuitively implies that it have a large amount of information. Likewise, the stability of the species implies a high degree of accuracy in transfer of this information. The cell, due to its finite size and lifetime, has some upper limit to the rate that it can order the atoms of the presumptive daughter cell by chemical reactions and physical associations. These aspects of cell division have been recast below into the language of information theory in order to place quantitative limits upon them. Once this is done, it is then possible to ask, by means of an uncertainty relation involving information and number of observations, whether it is possible to achieve the required accuracy of information transmission by determinate means available to the cell.[4]

[4] The two aspects, amount of complexity or information in a living cell and accuracy required for viable reproduction, which we join in a single calculation, may be employed individually to make a case for an organizing principle in biology. (See Elsasser, 1958 Chapter I.) In other words, each aspect may be utilized to calculate the great improbability of living organisms, which implies an organizing principle, since organisms do live, i.e. are in-existence. However, this does not permit a connection with physics. By combining the two aspects in a single calculation, we believe that it has been possible to keep a connection with physics and prevent a dualism in the theoretical background. No new organizing principle needs to be postulated.

An exact numerical estimate of the quantity of information transferred from cell to cell is not available and would be extremely difficult to obtain. Nevertheless, estimates of the upper and lower bounds are easily made. Dancoff and Quastler (1953) state that "the essential complexity of a single cell and of a whole man are both not more than 10^{12} nor less than 10^{5} bits". Table 1 lists some of the estimates along with the assumptions used as their basis. There is a remarkable concurrence in the value of the upper bound found for different organisms and by different means. The lower bound, which will be used in keeping with the conservative nature of the following calculation, is derived by Dancoff and Quastler. They consider each gene to be an independent source of information with an entropy which depends on the number of allelic states, i.e. different, recognizable messages on each gene. The information of the gene "catalogue" is taken to be the sole source of information for the cell, and is given by the number of genes times the information on each gene. (In the more recent terminology of molecular genetics (Monod, Jacob, and Gross, 1961), Dancoff and Quastler's "gene" would be termed an *operon* and probably include the regulator gene as well).

Though this lower bound will be used here, there is good reason to think that the actual essential information measure for the cell is nearer the upper bound. Linchitz points out that there is very little difference in heat output per organism between a rapidly or slowly growing culture; this indicates a relatively constant "heat maintenance." Even allowing for 99 per cent inefficiency in conversion of free energy into information (or negentropy) there would still be information content of 10^{10}–10^{11} bits.

The mutation rate may be determined experimentally for a given genetic character. For bacteria, these rates vary as is shown in Table 2. The nature of the calculation to follow requires an average mutation rate. This could be somewhere around 10^{-6} per bacterium per division. Again, in keeping with the conservative nature of the calculation the fastest mutation rate, 10^{-2} per bacterium, is chosen to represent the average accuracy of transfer for any given genetic character. This is to say, for any given genetic character there is a probability of 10^{-2} that it will be inaccurately transmitted. The other datum needed before proceeding to the uncertainty demonstration is the upper limit to the rate of information processing in the cell. Bremerman (1962) has shown that no

Table 1. Estimates of information in various organisms

Organism	Information (bits)	Investigator	Method of Estimation	Assumption of Methods
Average bacterium (size of *E. coli*)	1.3×10^{12}	Morowitz (1955) (*also* Setlow and Pollard (1962))	Number of bits required to specify kind and position of atoms in a cell, exclusive of water.	Interchange of any atom gives rise to a new organism or is fatal.
B. pycnoticus	4×10^{12}	Linschitz (1953)	All the free energy of metabolism is used to lower the entropy of the organism.	100% efficiency for metabolism and no energy used for other puposes, i.e. work of maintenance or osmotic regulation, etc.
E. coli	5.6×10^{11}	Morowitz (1955) (Data of Bayne-Jones and Rhees)	Direct measurement of heat associated with the production of one micro-organism.	(Same as immediately above.)
Average bacterium	10^5	Dancoff and Quastler (1953)	Information content in terms of number of genes and information contained on each.	A required developmental scheme, i.e. nothing novel supplied by the cytoplasm.
Human (germ cell)	10^7–10^9	Elsasser (1958)	Conservative estimate of information required to specify gross anatomy to within 1 mm^3.	Does not include the information required to specify the molecular structure and function of the germ cell itself.

Table 2. Mutation rates in bacteria (from Lamanna and Mallette, 1965)

Class of character	Mutation Rate (per bacterium per division)
Sensitivity to antibiotics	$\sim 10^{-10}$
Nutritional requirements	$\sim 10^{-10}$
Fermentation reaction	$10^{-5} - 10^{-7}$
Antigenic structure of flagella (phase variation)	$10^{-3} - 10^{-4}$
Colonial Morphology	$10^{-2} - 10^{-10}$
Pigmentation	$10^{-3} - 10^{-7}$

computer made of matter, and therefore subject to the mass-energy relation and to Heisenberg uncertainty, can possibly process more than 1.4×10^{47} bits of information per gram per second. This rate, of course, represents an upper limit for the cell, but it is an unrealistic one, since it presupposes an energy source that the cell does not have viz., matter-energy conversion. While even this rate will be shown to be incompatible with the minimum uncertainty in the required number of observations, it will be somewhat more satisfactory to demonstrate a narrower limit for the information transfer rate of the cell. By doing so, the assertion that information transfer is indeterminate at the molecular level will be justified. An estimate of the upper limit for the total number of possible observations during the lifetime of an *E. coli* may be made by assuming that each of the 10^{10} atoms contribute to $3\,n - 6$ vibrational modes. Each mode now represents one half of a possible information transfer channel operating at an upper limit of the vibrational frequency of 10^{15} bits per second. This assumption, in effect, takes the cell to be one giant molecule, by converting the translational modes into vibrational modes. This occasions no loss of generality and at the same time assures that all potential chemical reactions (change of vibrational modes of activated complex into translational modes of products [Glasstone, 1946] are included in the numbering scheme. The lifetime of an *E. coli* on optimal medium is about 20 min. (Setlow and Pollard, 1962, p. 499). These numbers give an upper limit to the information transfer of the typical bacterium having 10^{10} atoms of $3/2 \times 10^{10} \times 10^{15} \times 10^{3} = 10^{28}$ bits/lifetime. (To see how this number compares to that given by Bremerman, we need only to increase the frequency to that represented by the nuclear binding energies, 10^{23}–10^{25}/sec, and by the number of *E. coli* per gram, 10^{12}, obtaining 10^{45}–10^{47} bits/gram/second.)

To demonstrate the complementary nature of the variables of an experiment requiring the simultaneous measurement of information, H, and the number of observations, N, there needs to be shown that $\Delta H \cdot \Delta N = C$, where C is a constant and Δ represents a suitable measure of deviation (Elsasser, 1960 [appendix of article]). Starting with the normalized error curve, the probability, p, is given by

$$p = \frac{1}{\sqrt{2\pi}} e^{-\frac{x^2}{2}}$$

with x measured in σ units from the mean. By taking logarithm of both sides, there results

$$ln\ p = ln \frac{1}{\sqrt{2\pi}} - \frac{x^2}{2}.$$

The equation, with variables $ln\ p$ and x^2, represents a straight line to the right of the ordinate with an intercept, $ln \frac{1}{\sqrt{2\pi}}$ and a slope, $-1/2$. (See Figure 5). The area in the first quadrant is

$$\left(ln \frac{1}{\sqrt{2\pi}}\right)^2,$$

reflecting the normalization.

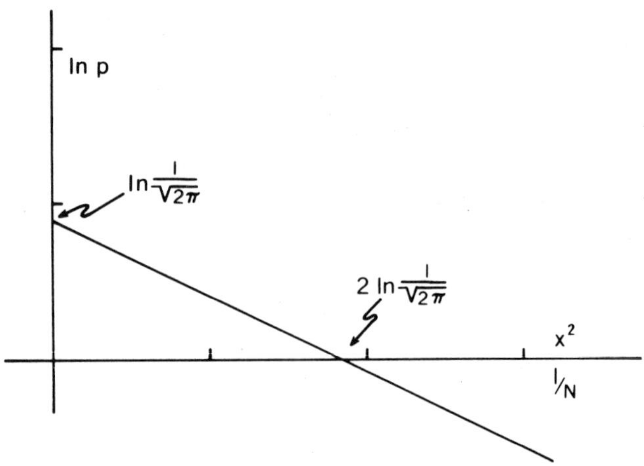

Figure 5. The Gaussian distribution with coordinates x² *and* ln p.

Defining the differences Δ between any two values of the variables

$$\Delta\, ln\, p = ln\, p_2 - ln\, p_1$$

$$\Delta\, ln\, x^2 = x_2^2 - x_1^2 \quad ,$$

the two point equation for a line gives

$$\Delta\, ln\, p = -1/2\, \Delta\, x^2$$

or

$$\Delta\, (-\, ln\, p) = 1/2\, \Delta\, x^2 \quad .$$

For a Gaussian distribution, σ is the deviation about the mean and is inversely proportional to the square root of the number of observations, giving

$$\sigma^2 = \alpha/N \quad .$$

Also, by definition

$$x^2 = \beta\sigma^2$$

$$x^2 = C/N$$

with α, β, and C various constants of proportionality. Since by definition

$$\Delta\, (1/N) = 1/N_2 - 1/N_1 = \frac{N_1 - N_2}{N_2 N_1} = \frac{\Delta\, N}{N_2 N_1}$$

for sufficiently large N, that is, $1/N$ near the origin, $N_2 N_1$ is approximately constant, and we find that

$$\Delta\, (1/N) = \gamma \Delta N$$

with γ another constant of proportionality. Then

$$\Delta\, x^2 = C'/\Delta N$$

and

$$\Delta - ln\, p \cdot \Delta N = C'/2$$

which demonstrates the complementary nature of the variables.

To determine the value of C', substitute $x^2 = C'/N$ into the original equation, obtaining

$$ln\, p = ln \frac{1}{\sqrt{2\pi}} - \frac{C'}{2}\left(\frac{1}{N}\right),$$

which represents a change of scale of the abscissa. As N increases, the probability becomes a maximum, reflecting the concentration of the distribution according to the law of large numbers (Eisenschitz, 1958, pp. 77–8).

At the point $ln\ p = 0$, $1/N$ must be

$$2\ ln \frac{1}{\sqrt{2\pi}}$$

to preserve normality; therefore, $C' = 1$ and $\Delta - ln\ p \cdot \Delta N = 1/2$. Linschitz (1953) gives the relation between information and probability with equal expectation: $H\ ln\ 2 = ln\ P$, P represents the number of possible ways an event may be realized. P is related to p, the probability of an event, by $P = 1/p$ for events of equal probability (Setlow and Pollard, 1962, p. 67). Since $ln\ P = ln\ 1/p = -ln\ p = H$, the desired relation becomes $\Delta H \cdot \Delta N = 1/2ln\ 2 \approx 0.72$, which is the needed relation.

$\Delta ln\ p$ may be termed the requisite accuracy of information, and is related both to the amount of information, given by $\Delta ln\ p/ln\ 2$ and the observed mutation rate. The mutation rate represents the average deviation, Δp, about any character whose information content is related to the p above.
By definition

$$\Delta ln\ p = |\ ln\ p'' - ln\ p'\ | = ln \frac{p''}{p'}\ ,$$

where p'' and p' represent the upper and lower value of the deviation about p. (See Figure 6). (By taking care to choose $p'' > p', p''/p' > 1$ and

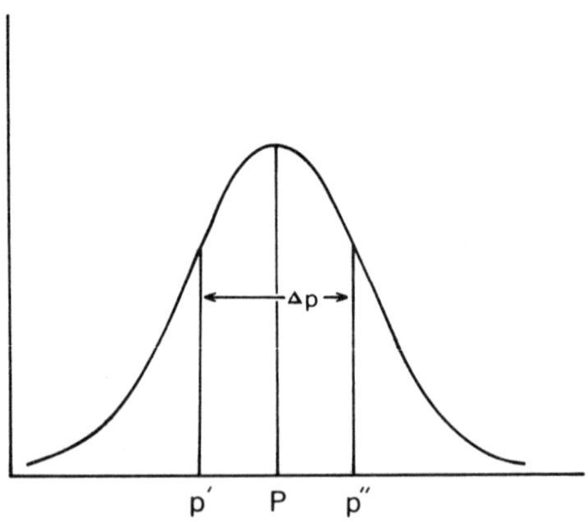

Figure 6. Distribution about p.

$ln\ (p''/p') > 0$, the absolute value sign can be dropped.) The problem of determination of $\Delta ln\ p$ can be stated in the following form.

Given:

$$\Delta p = p'' - p' \tag{1}$$

and the value of p, find p''/p' which will give directly

$$ln\left(\frac{p''}{p'}\right) = \Delta ln\ p.$$

By the symmetry of the distribution

$$2p = p'' + p'. \tag{2}$$

Substitute (1) in terms of p' into (2), obtaining

$$2p = 2p'' - \Delta p. \tag{3}$$

Again substituting (1) in terms of p'' into (2)

$$2p = 2p' + \Delta p \tag{4}$$

upon rearranging (3) and (4) and dividing, we obtain

$$\frac{p''}{p'} = \frac{2p + \Delta p}{2p - \Delta p}$$

with

$$ln\left(\frac{p''}{p'}\right) = \Delta ln\ p = ln\left(\frac{2p + \Delta p}{2p - \Delta p}\right),$$

which is the desired expression.

According to the relation given by Linschitz

$$H ln 2 = ln\ p$$

with $H = 10^5$, p becomes

$$p \approx (e)^{10^5} \approx (10)^{10^4},$$

and in the expression for $\Delta ln\ p$, p is scarcely affected by the addition or subtraction of the small number Δp. This is to say, the distribution is very concentrated about p. The ratio p''/p' is very close to one, while its logarithm is correspondingly close to zero. The uncertainty in ΔN becomes thereby a very much larger number.

To show this more precisely, the expression for $\Delta ln\, p_0$ will be expanded in a logarithmic series of the type:

$$ln\,(a + x) = ln\, a + 2\left[\frac{x}{2a + x} + \frac{1}{3}\left(\frac{x}{2a + x}\right)^3 + \frac{1}{5}\left(\frac{x}{2a + x}\right)^5\right] + \ldots$$

in the interval

$$a > o,\ -a < x < +\ \infty \quad .$$

Upon setting $x = \Delta p_0$ and $a = p_0$ we obtain, ignoring the higher terms of the expansion:

$$\Delta ln\, p_0 \sim ln\, p_0 + 2\left(\frac{\Delta p_0}{2p_0 + \Delta p_0}\right) - \left[ln\, p_0 + 2\left(\frac{-\Delta p_0}{2p_0 - \Delta p_0}\right)\right].$$

With $p_0 \gg \Delta p_0$ we may also ignore the Δp_0 terms of the denominator to obtain the simple expression:

$$\Delta ln\, p_0 \sim 2\frac{\Delta p_0}{p_0} \sim 2\frac{10^{-2}}{(10)^{10^4}} \sim (10)^{-10^4} \quad ,$$

which by the uncertainty relation gives

$$\Delta N \sim (10)^{10^4} \quad .$$

This minimum value for the uncertainty is immensely larger than the upper limit for the number of molecular observations possible during the life time of the cell. Each offspring of the cell would have a vanishingly small chance for viability under a mechanistically determinate basis for information transfer. In other words, *it is impossible to follow observationally the molecular motions associated with any arbitrarily chosen genetic character during its transmission from cell to cell.* This is such a serious proscription in the light of current projections of molecular genetics that some comment must be made concerning what is not said. First of all, the majority of the information is transferred indeterminately; certain characters may be followed at a molecular level, but these will be relatively few when compared to the entire genome. This uncertainty also does not imply that certain morphological aspects of the cell, e.g. chromosome bands or nucleotide sequences, may not be *associated* with particular genetic characters. The association will be empirical and will not generally be amenable to a continued analysis in terms of molecular motion. Several lines of experimental evidence point in this direction.

Davis and associates (1965) have shown that a variety of small and ubiquitious molecules can cause ambiquities in DNA transcription in *in vitro* systems. Marshak and Marshak (1954) give evidence of the complete disappearance of DNA from the echinoderm egg. Amitotic division, lacking a mechanism to assure even division of the chromosome complement, may occur in stages of germ cell proliferation (Child, 1907) or in solitary organisms (Nathanshon, 1900).

Viruses or transforming factors, having a small genome, <50 characters, may be capable of almost complete analysis in terms of information transmitted, i.e. in terms of replication or cellular changes. It must be remembered that an infected or transformed cell supplies all of the machinery necessary to accomplish the "function" of the particle.

In the face of this molecular indeterminacy inherent in information transfer, how is the known information and accuracy to be accounted for? Is this a paradox? In a sense, the first question is analogous to asking, "What keeps the electron from spiraling into the atomic nucleus?" This does not, however, satisfactorily answer the question nor resolve the paradox. The resolution will be accomplished in the next chapter where it is shown that systems of molecules of such nature as to prohibit the use of the central limit theorem in describing their statistics, perforce, must cyclically reproduce systems with like states. This in no way circumvents the molecular indeterminacy, since Gaussian distributions —as assumed here—lead to the minimum uncertainty product.

The indeterminacy principle is a means for testing for *external* consistency of a mechanistically determinate model with respect to other mechanistically determinate models in the same system. Historically, the most basic scientific description of a process has been in terms of a mechanistic model. The experimental arrangements for study of biological processes have usually been carried out so a single variable of motion can be measured with precision. The experimental arrangements have then been altered to measure some other mechanical variable. In this manner the *internal* consistency of a proposed mechanistically determinate model has been subjected to experimental verification. The external consistency of the model has been assumed in most instances, that is, the assumption has been made that the model will be consistent with models for other processes in the same cell or organism. In other words, most biological investigations have been conducted with the tacit assumption that inconsistencies between mechanistically determinate models in the same system will disappear at the molecular level. However, at the level of molecules the indeterminacy principle can not be disregarded. Here, the indeterminacy principle is not only useful in indicating whether a given model has the possibility of being proved or disproved, but this principle has the positive meaning that complementary modes must be utilized in making descriptions of mechanistically indeterminate processes, if the tests of external as well as internal consistency are to be satisfied. This complementarity of wave and particle description of atomic processes has been amply demonstrated. The implication of this monograph is that complementarity must not be overlooked in the description of many living processes.

References

Ashby, W. R. 1963. *An Introduction to Cybernetics*. New York: Wiley.

Barnothy, J. M. 1960. In *Medical Physics*, ed. O. Glasser, pp. 61–64. Chicago: Year Book Medical Publishers.

Bohr, N. 1961. *Atomic Physics and Human Knowledge*. New York: Science Editions.

Botts, J. 1957. In *Physiological Triggers*, ed. T. H. Bullock, pp. 85–102. Washington: *Am. Physiol. Soc.*

Bremerman, H. J. 1962. In *Self-Organizing Systems*, eds. M. C. Yovits, G. T. Jacobi, and G. D. Goldstein, pp. 93–106. Washington: Spartan Books.

Brillouin, L. 1962. 2d ed. *Science and Information Theory*, New York: Academic Press.

Brown, F. A., Jr. 1959. *Science* **130**: 1539.

Chambers, R., and Chambers, E. L. 1961. *Explorations into the Nature of the Living Cell*. Cambridge: Harvard University Press.

Child, C. M. 1907. *Biol. Bull.* **12**: 89.

———. 1907*a*, *Biol. Bull.* **12**: 179.

Dancoff, S. M., and Quastler, H. 1953. In *Information Theory in Biology*, ed. H. Quastler, pp. 263–73. Urbana: University of Illinois Press.

Davidson, E. H. 1965. *Scientific American* **212**: 36.

Davis, E. W., So, A. G., Bodley, J., and Weinberg, F. 1965. *Science* **150**: 371 (abstract).

Eagle, H. 1959. *Science* **130**: 432.

Eisenschitz, R. 1958. *Statistical Theory of Irreversible Processes*. London: Oxford University Press.

Elsasser, W. M. 1958. *The Physical Foundation of Biology*, New York: Pergamon.

Elsasser, W. M. 1960. *J. Theoret. Biology* **1**: 27 (see appendix of this article).

Fatt, P., and Katz, B. 1953. *Acta Physiol. Scand.* **29**: 117.

Feldman, I. 1956. *Analytical Chemistry* **28**: 1859.

Girardier, L., Ruben, J. P., Brandt, P. W., and Grundfest, H. 1963. *J. Gen. Physiol.* **47**: 189.

Glasstone, S. 1946. 2d ed. *Textbook of Physical Chemistry*. Princeton, N. J.: Van Nostrand.

Grisolia S. 1964. *Physiol. Revs.* **44**: 657.

Heilbrunn, L. V. 1943. 2d ed. *An Outline of General Physiology.* Philadelphia: Saunders.

Heilbrunn, L. V. 1956. *The Dynamics of Living Protoplasm,* New York: Academic Press.

Hoch, F. L., and Vallee, B. L. 1952. *J. Biol. Chem.* **195**: 531.

Hoffman, J. G. 1957. *The Life and Death of Cells,* Garden City, N.Y.: Doubleday.

Hogness, D. S. 1959. In *Biophysical Science,* ed. J. L. Oncley, pp. 256–58. New York: Wiley.

Huxley, A. F., and Taylor, R. E. 1958. *J. Physiol.* **144**: 426.

King, T. P., Norman, P. S., and Connell, J. T. 1964. *Biochemistry* **3**: 458.

Klotz, I. M., Ayers, J., Ho, J. Y. C., Horowitz, M. C., and Heiney, P. E. 1958. *J. Am. Chem. Soc.* **80**: 2132.

Koshland, D. E., Jr. 1959. In *The Enzymes,* eds. P. D. Boyer, H. Lardy, K. Myrback, Vol. I, pp. 305–46. New York: Academic Press.

Lamanna, C., and Mallette, M. F. 1965. 3d ed. *Basic Bacteriology.* Baltimore: Williams and Wilkins.

Lamanna, C. 1959. *Science* **130**: 763.

Linschitz, H. 1953. In *Information Theory in Biology,* ed. H. Quastler, pp. 251–62. Urbana: University of Illinois Press.

Lumry, R. 1959. In *The Enzymes,* eds. P. D. Boyer, H. Lardy, K. Myrback, Vol. I, pp. 157–231. New York: Academic Press.

Lwoff, A. 1966. *Science* **152**: 1216.

Mackay, S. 1960. *I.R.E. Transactions ME–7*: 94.

Marshak, A., and Marshak, C. 1954. *Nature* **174**: 919.

Martin, A. R. 1966. *Physiol. Rev.* **46**: 51.

McIlwain, H. 1946. *Nature* **158**: 988.

Monod, J., Jacob, F., and Gross, F. 1961. *Biochem. Soc. Symp.* **21**: 104.

Morowitz, H. 1955. *Bull. Math. Biophys.* **17**: 81.

Nathanshon, A. 1900. *Jahrb. Wiss. Bot.* **35**: 1.

Perry, S. V. 1956. *Physiol. Revs.* **36**: 1.

Polanski, J. 1964. In *Electronic Aspects of Biochemistry,* ed. B. Pullman, pp. 481–502. New York: Academic Press.

Porter, K. R., and Palade, G. E. 1957. *J. Biophys. Biochem. Cyto.* **3**: 269.

Puck, T. T., and Marcus, P. I. 1956. *J. Exptl. Med.* **103**: 653.

Rahn, O. 1933. *Invisible Radiations of Organisms, Protoplasma—Monographien,* Vol. 9. Berlin: Verlag von Gebruder Borntraeger.

Rayns, D. G., Simpson, F. O., and Bertand, W. S. 1967. *Science* **156**: 656.

Schoenheimer, R. 1942. *The Dynamic State of Body Constituents.* Cambridge: Harvard University Press.

Setlow, R. B., and Pollard, E. C. 1962. *Molecular Biophysics.* Reading, Mass. Addison-Wesley.

Sollberger, A. 1962. In "Rhythmic Functions in the Living System," ed. W. Wolf, *Ann. N.Y. Acad. Sci.* **98**: 757.

Sonnenblick, E. H. 1962. *Federation Proceedings* **21**: 975.

Spector, W. S., ed., 1956. *Handbook of Biological Data.* Philadelphia: Saunders.

Stetten, D., Jr. 1959. *In Biophysical Science,* ed. J. L. Oncley, p. 563. New York: Wiley.

Thompson, D. W. 1942. *Growth and Form*, Vol. 1. London: Cambridge University Press.

Wald, G. 1965. *Science* **150**: 1028.

Wilder, J. 1962. In "Rhythmic Functions in the Living System," ed. W. Wolf, *Ann. N.Y. Acad. Sci.* **98** (art. 4): 1211.

Wilkie, D. R. 1960. In *Progress in Biophysics and Biophysical Chemistry*, Vol. 10, eds. J. A. V. Butler and B. Katz, pp. 259–98. New York: Pergamon.

CHAPTER 4

TRANSFORMATIONS AS QUANTUM RULES

The forms of classes of transformations as quantum rules for essential processes in living units based on phenomenological conditions under which these processes occur

From this point on, the argument proceeds on the basis that it has been plausibly established (*a*) that there are two general, essential classes of physical processes in living units (bions) which have a *mechanistically indeterminate* basis of operation by virtue of the phenomenological conditions of their molecular motion which preclude assumption of Gaussian distributions of conjugate variables; (*b*) that descriptions of these classes of processes in living units must, therefore, have a *quantal* or discrete feature; (*c*) that the title, *emergent*, may be used to name these quantized processes; and (*d*) that the quantum description of these emergent processes is *nondualistic* with present physical theory under a correspondence rule. This chapter will further qualify the quantum description which is applicable to these emergent processes.

The trajectory of any system in which there are distinguishable energy states is describable with transformations (Ashby, 1963, Chap. 2). Mechanical theory is the most general means of specification of these transformations. These transformations are expressed mathematically as matrices. By the use of mathematical theory, various kinds of matrix algebras may be employed in an attempt to represent the transformations in correspondence with characteristics of physical processes of a system. Two pathways are available to explore the consequences of some kind of matrix algebra in the evolution of a mathematical method in order to limit the classes of matrices that are involved in describing biological processes. The first pathway is to set up a series of assumptions of the kind of algebra through which the matrices are going to be manipulated, see what the consequences are of manipulating the matrices in this fashion, and show that these consequences correspond to some new or old or unexpected way of viewing biological processes. This pathway does not allow the intuition which is necessary for arriving at the proper algebra; that is, there is no way of achieving correspondence of the physical characteristics of the biological processes with reality. The second pathway requires that in the act of setting up the algebra, an attempt must be made to keep some kind of correspondence with physical characteristics of the processes themselves. This is necessary if the axiomatization of the new algebra is to correspond with the characteristics of the new processes that the algebra is to cover. Neither of the above mathematical pathways is followed in this monograph. Recourse is taken in the direction of specifying the forms of the classes of transformations by means of the phenomenological conditions which engender indeterminacy.

The latter course is adopted in view of the facts that the most general form of the indeterminacy relation can be derived from the quantum postulates and that these postulates must be applicable by means of a mechanics which is capable of describing processes whose indeterminacy depends upon phenomenological conditions that preclude attainment of Gaussian distributions of conjugate variables.

Actual solutions of the mechanical equations are not possible since, for essential classes of processes in living units, there is no obvious way of appropriately assuming some specific matrix density of all points of momentum and position, or energy and time, in unitary space of the state function (which is constructed for obtaining discrete or quantal solutions in mechanical equations for the processes). Even though description in terms of magnitude is not possible, description in terms of relation is possible in the light of the known nature of reproductive and metabolic processes in living units. Teleonomically (Mayer, 1961), transformations may be constructed as rules for the proper forms that the solutions do take. (The transformations for these proper forms of the solutions constitute quantum rules which should be very useful in biology, as they are in chemistry.) For the essential classes of processes in living units, the nature of the classes of transformations which are constructed as quantum rules must be isomorphic with the most general phenomenological conditions of the processes.

SPECIFICATION OF THE NATURE OF THE CLASSES OF TRANSFORMATIONS BY ISOMORPHISM WITH PHENOMENOLOGICAL CONDITIONS OF ESSENTIAL CLASSES OF PROCESSES IN LIVING UNITS

The nature of the classes of transformations may be specified by means of the characteristics, or most general phenomenological conditions, which have been established for the essential classes of processes that occur in living units, namely, (*a*) discreteness of action, (*b*) radically limited number of occupied energy states in reproductive processes, and (*c*) radically large number of energy states accessible for occupancy in metabolic processes.

One specification of a process, if it must be represented as being indeterminate depending on an insufficiently large number of occupied energy states, is that the process must be radically recurrent or limited in number of molecules in time and space and, hence, be *cyclic* in nature.

Another specification of a process, if it must be represented as being indeterminate depending on an immensely large number of energy states which are accessible for occupancy, is that the process must be radically open in molecular mass and energy transport with environment and, hence, be *asymmetric*, *irreversible*, *and of a nonequilibrium nature.*

These specifications, along with the specification of quantization or discreteness of action, are sufficient to begin to describe the classes of transformations. The forms of these classes of transformations will be kept very simple purposely, in order not to pretend more than can be accomplished at present. One of the legitimate functions of science is to observe phenomenological conditions. In the future, as these conditions are discovered, it may be possible to elaborate upon the forms of the transformations.

The general nature of these classes of transformations is not unlike the general nature of classes of cybernetic transformations.

TRANSFORMATIONS APROPOS OF CYBERNETICS

Ashby (*loc. cit.*) has stated that cybernetics is a "theory of machines," but that it treats ways of behaving, rather than things. It asks what a thing does, not what a thing is. Cybernetics deals with any behavior that is regular or reproducible. It is not dependent on properties of matter or laws of physics. The advantage of cybernetics is that it deals with the possible relations between "machine" and "machine." "Machine" in this sense is relieved of its deterministic implication.

The classes of transformations which are of interest in this monograph and which are, in general, similar to those of cybernetics can be considered here briefly and in an elementary manner, as is compatible with our purpose.

In a transformation an operator can act upon a set of operands which undergo transition to a set of transforms. A good example is that of coding which requires each letter of a message to become the next letter in the alphabet. A becomes B, B becomes C, ... Y becomes Z, Z becomes A. No causality is inferred in the transformation. A statement is merely made of a set of operands and what each is changed to.

When no new transform appears which is not already among the operands, the set of operands is closed under the transformation. Closure is a relation between a particular set of operands and a transformation. Causality, again, is not referred to; the details of the transformation itself are the criteria for closure.

Example I of a transformation which undergoes closure:

$$\downarrow \begin{matrix} 1 & 2 & 3 & 4 \\ 2 & 3 & 4 & 1 \end{matrix}$$

Example II of a transformation which is not closed:

$$\downarrow \begin{matrix} \mathrm{A} & \mathrm{B} & \mathrm{C} & \mathrm{D} \\ \mathrm{E} & \mathrm{G} & \mathrm{H} & \mathrm{I} \end{matrix}$$

Repeated action of a transformation requires that the transformation be closed. When a transformation is applied repeatedly, a series of changes is generated similar to a dynamic system. From each closed transformation another closed transformation is obtained, the effect of which applied once is the same as the original transformation applied two times or three times or four times and so on, as "powers" of the original transformation. Ashby has called the original transformation T and subsequent powers T^2, T^3, T^4, and so on.

Transformations are single-valued if they change each operand to only one transform. Examples III and IV are single-valued transformations which undergo closure:

$$\downarrow\begin{matrix}1 & 2 & 3 & 4\\ 2 & 1 & 1 & 4\end{matrix} \quad \text{or} \quad \downarrow\begin{matrix}1 & 2 & 3 & 4\\ 2 & 3 & 4 & 1\end{matrix}$$

Example V is a non-single-valued transformation which undergoes closure:

$$\downarrow\begin{matrix}1 & 2 & 3 & 4\\ 2 \text{ or } 4 & 1 & 2 \text{ or } 3 & 4\end{matrix}$$

Single-valued transformations are one-to-one, if all the transforms are different from one another. Example VI is a single-valued, one-to-one transformation which is closed:

$$\downarrow\begin{matrix}1 & 2 & 3 & 4\\ 2 & 3 & 4 & 1\end{matrix}$$

A single-valued transformation that is described as many-to-one does not have individual transforms which are unique. Example VII is a single-valued, many-to-one transformation which is closed:

$$\downarrow\begin{matrix}1 & 2 & 3 & 4\\ 2 & 1 & 1 & 4\end{matrix}$$

An identical transformation is one in which each transform is the same as its operand, i.e. no change occurs. Example VIII is an identical transformation:

$$\downarrow\begin{matrix}1 & 2 & 3 & 4 & 5 & 6\\ 1 & 2 & 3 & 4 & 5 & 6\end{matrix}$$

A transformation can be single-valued and one-to-one with addition of a constant interval. Example IX is as in an Archimedean spiral intersecting with a line:

$$n \longrightarrow n + 1 \; (n = 1,2,3,4, \ldots\ldots\ldots\ldots)$$

$$\downarrow\begin{matrix}1 & 2 & 3 & 4 & \ldots\ldots\ldots\ldots\\ 2 & 3 & 4 & 5 & \ldots\ldots\ldots\ldots\end{matrix}$$

A transformation can be single-valued and one-to-one with addition of a logarithmic interval. Example X is as in a logarithmic spiral intersecting with a line:

$2n \rightarrow 2(2n)$	$(n = 1,2,3,4, \ldots\ldots\ldots\ldots)$
2 4 8 16	32 ....
↓ 4 8 16 32	64 ....

Other transformations follow from the combination of two or more of these basic transformations. The Fibonacci series approximates the results of two processes operating simultaneously, one represented under an exponentially decreasing transformation and the other under a periodic transformation (with equal intervals).

Transformations may be constructed with parameters a,b,c,d, etc. The particular transformation to be accomplished is selected by choice of a specific parameter. Examples XI and XII:

	1	2	3	4		5	6	7	8
Xa	2	1	1	4	Ya	7	8	6	5
Xb	3	4	1	2	Yb	6	7	8	5
Xc	2	3	4	1	Yc	8	7	5	6
Xd	4	3	1	2	Yd	6	8	5	7

We return now to the specification of classes of transformations for classes of processes in biology. A note on the cyclic nature of biological processes will lead into the actual specification of classes of transformations for genetic and epigenetic information transport processes.

CYCLIC NATURE OF REPRODUCTIVE PROCESSES IN BIOLOGY

The hypothesis has been proposed in this manuscript that many biological processes which emerge at a given level depend on the indeterminate nature of the process at a lower level of organization, that is, at the level of large molecules. If this be true, one may look for an inherent cyclic behavior of many of these emergent processes. This recurrent behavior seems to be required for such processes which operate on a mechanistically indeterminate basis, since beyond a certain time interval the occupied energy states of the process would be sufficient in number for the assumption that the process approximates conventionally statistical or thermodynamic behavior. In other words, if a process be emergent on a basis of indeterminacy of microparticulate motion due to limitation of number of molecules, the process must demonstrate certain limitations or constraints. A most obvious way to this end is for the process to develop along lines of radically recurrent or cyclic behavior in which the count of accessible energy states in space and time passes through zero and must begin anew. This behavior characterizes the "reproductive" classes of processes. (Actually, one may not separate cause and effect in evolution so nicely as seems to have been done here. The purpose is merely to demonstrate the interdependence of the two.)

Two further comments should be made. The first is that this cyclic nature need not demonstrate an obvious rhythm. There may be only a temporal constraint or time envelope under which an emergent process may occur once or a few times or many times. In this manner the build-up of number of occupied energy states is restricted. This is associated with a necessarily fuzzy maximum time interval—i.e. one species or strain may tolerate a close approach phenomenologically (of a given emergent process) to conditions for which conventionally statistical or thermodynamic ensembles may be assumed, while another may not, even to the opposite extreme.

The second comment is that, without taking from the validity of the aforesaid, genetic and epigenetic processes might well have evolved strict temporal (clock-like) periodicities that do insure an indeterminate basis of operation of molecular motion. These may be inherent, in the sense of being emergent from within themselves, such as the well known circadian rhythms that persist in the absence of light and dark cycles. They may also have adapted to take their clues from regular and dependable changes in the external environment, such as the light-dark cycles, lunar tidal rhythms, seasonal variation of heat-cold, etc. (the list is very long). More complex organisms have internal variations such as hormonal outputs and sleep-wake cycles that give still other synchronizing systems from which various dependent cells of the organism take their temporal clues.

The length of some cycles immediately raises the question of whether a molecular process can take such a long time without constituting a phenomenological background which may be described under a statistical ensemble? There is no reason why not, especially when one is considering processes involving the action of a few macromolecules, for instance, the nucleic acids undergoing comparatively slow processes, such as replication.

Genetic and epigenetic processes might also have evolved requisite spatial regularities or periodicities that preclude the build-up of sufficient number of occupied energy states, and insure an indeterminate basis of operation of molecular motion. Just as with temporal regularities, the underlying basis of space-limiting may itself be emergent, may take various clues from external and/or internal environment, and may or may not be very obvious as to strict regularities.

Finally, some emergent processes may be admixtures of both time and space limitation. No immediate logical reason requires that in selected instances temporal constraints may not be traded off against spatial constraints, completely obscuring regularities which might have been observed only in one mode.

In summary, it seems that information conservation such as that in epigenetic and genetic processes may exhibit time, space, and time-space regularities which do insure a larger indeterminacy of molecular motion. However, the converse need not be true. Every temporal rhythm or spatial periodicity does not necessarily exist phenomenologically in order to insure a larger indeterminacy which includes molecular motion.

SPECIFICATION OF THE FORMS OF CLASSES OF TRANSFORMATIONS WHICH REPRESENT THE DISCRETE, CYCLIC NATURE OF REPRODUCTIVE PROCESSES

Series of positive integers that will remain invariant under classes or sets of transformations may be employed to represent the discrete, cyclic nature of the reproductive class of processes.

These series of positive integers may duplicate themselves with various intervals. There are several possible types: duplicating series with identical intervals; duplicating series with addition of linear or equidistant intervals, as illustrated by points of intersection of a nonlogarithmic (Archimedean) spiral with a line; duplicating series with addition of logarithmic intervals, as illustrated by points of intersection of a logarithmic spiral with a line; and duplicating series with addition of combinations of linear and logarithmic intervals (such as the Fibonacci series). The positive integer series will be composed of limited numbers of integers to stay outside the classical limit imposed by a correspondence rule. The general statement may be made that the transformations are single-valued and one-to-one and will undergo closure, meaning that the transformation will be continually applicable. See examples VI, VIII, IX, and X on pages 71, and 72.

Thus we have a lengthy but complete chain of reasoning leading to the continuity of biological information transport (or conservation of information in reproduction) on a mechanistically indeterminate basis for which discrete, cyclic classes of transformations have been specified.

This aspect may be summarized by stating that a process must be of such a nature that it will duplicate itself under a required *cyclic* operation, if it has a radically insufficient number of occupied energy states and an enlarged uncertainty product in an uncertainty relation, so that the process emerges as indeterminate, hence, discrete (quantized). The forms of this class of reproductive processes may be represented by quantum rules under positive, duplicating integer series in single-valued transformations which reflect the cyclicity that is necessary to restrict radically the number of molecules in time and space. (The latter restriction is the most general phenomenological condition of the class of reproductive processes.) Note that this information conserving type of process is essentially closed or does not exchange mass and energy with environment.

SPECIFICATION OF THE FORMS OF THE CLASS OF TRANSFORMATIONS WHICH REPRESENT THE DISCRETE, ASYMMETRIC NATURE OF METABOLIC PROCESSES

A process must be of such a nature that it will be open or irreversible and nonequilibrium under a required *asymmetric* operation, if the process has a radically large number of energy states which are accessible for occupancy and if the process has an enlarged uncertainty product, such that it emerges as indeterminate, hence, discrete (quantized). The form of this class of metabolic processes may be represented by quantum rules under positive integer series duplicating in multivalued transformations which reflect the *asymmetry* that is necessary to keep radically open the transport of mass and energy with environment (the latter restriction is the most general phenomenological condition of the class of metabolic processes). The general statement may be made that this class of transformations is non-singled-valued (multivalued), one-to-one or many-to-one, and will undergo closure, meaning that the transformations will be continually applicable. See example V on page 71.

CLASSES OF TRANSFORMATIONS AND GENERAL BIOLOGICAL LAWS

The general biological laws which are presently recognized may be reviewed in order to evaluate whether these laws can be interpreted as quantum rules. The criteria for an affirmative answer may be formulated in terms of the form and constitution of quantum rules in each situation. Quantum rules here are statements concerning the nature of classes of transformations for representation of the proper forms that solutions of mechanical equations do take under the general phenomenological conditions of the essential classes of processes of living units. Accordingly, for affirming that a particular general biological law is in effect a quantum rule, the criteria are: (*a*) the class of processes to which the general biological law refers is one of the essential ("metabolic" and "reproductive") classes of processes of living units, and (*b*) the nature of the classes of transformations which is implied in the general biological law is isomorphic with the general phenomenological conditions under which the class of processes occur.

Two general biological laws which are known (and which seem to be only two among a group of similar laws) are: the *Law of Initial Value* of physiology and the *Arndt-Schulz Law* of pharmacology. These laws refer to the "metabolic" class of processes, and they satisfy the required isomorphism for quantum rules. By way of discussion, this group of laws might be described as involving a stimulus or a dose-response or intensity-sensation type of process, with the characteristic that the resulting response is not simply proportionate to the stimulus or dose. Another energy input or power source is required, which means specifically that there is a multivalued nature in this type of operation. In other words, there seems to be a number of processes which act as if they were operating with a "trigger" feature, or as if they were comprised of "incremental negative resistance" elements. A very general set of biological laws of metabolism (like the set of laws of inheritance) is suggested here to cover this kind of lawfulness which occurs under a dose-response or intensity-sensation relation in living units. A logarithmic function or linear relation (Fechner's law) can be found over short segments of the graphic curve which represents dose-response or intensity-sensation relations in neurological testing, for example, or in the stimulating effects of low doses of anesthetics, or in Starling's law of the heart.[1] A power law is discovered upon examination of the extreme values of the curve. This is particularly true at the region where reversal of response is approached. This situation, including the paradoxical response, has the operational characteristics of "incremental negative resistance" elements. This subject is discussed further in Chapter 9 with reference to irritability, contractility, homeostasis, and temporal rhythms and spatial patterns of metabolism. For the present, it is sufficient to say that there is an elegant isomorphism between the multivalued nature of the class of "incremental negative resistance" operations and the multivalued nature of the class of transformations under which metabolic processes have been subsumed above.

The only other general biological laws which are known refer to the reproductive class of processes and are (in the incomplete form that we know them): the First and Second Mendelian Laws of Inheritance in genetics (along with other phenomenological factors of heritable variability, such as partial genetic blocks and somatic mutations), and the Laws of Cellular and Tissue Differentation in epigenesis (ontogeny recapitulates phylogeny, for instance). There is an elegant isomorphism between the single-valued nature of the *bionic* class of gene-duplicating operations and the single-valued nature of the class of transformations that duplicate themselves with identical or linear intervals, under which genetic processes have been subsumed above. Likewise, there is an elegant isomorphism between the single-valued nature of the *cellular* class of gene duplicating operations and the single-valued nature of the class of transformations which duplicate themselves with logarithmic or combined logarithmic and linear intervals, under which epigenetic processes have been subsumed above.

OVERT INSTANCES WHERE NEW QUANTUM RULES OBTAIN IN BIOLOGY

There are overt occasions when the application of quantum theory for processes with larger indeterminacy will be seen to obtain in biology in the form of new quantum mechanical rules. As an example, the gen-

[1] Starling's law of the heart states that the contractile action of the heart is related to the stretch of the heart muscle at the commencement of contraction. A "descending limb" occurs ultimately.

eralization was cited that the condition, insufficient number of occupied energy states in a process, meaning the distribution of conjugate variables is not assumable as Gaussian, is *necessary* (even if not sufficient) before reproduction of a new organism or biological unit occurs. All organisms require restriction upon the number of molecules in time and space in order to transfer genetic information. Even the slime mold, which seems to exist in an exceptional manner as an amorphous, syncytial form, changes into cellular units, which facilitates reproduction (with the formation of fruiting bodies). This has been deduced from the form that one class of the quantum rules has to take in the quantum mechanics which involves processes with larger indeterminacy. The form of this class of quantum rules is: the class of genetic information conservation processes obey a class of transformations with a closed cyclic (single-valued) nature. The phenomenological condition for the closed cyclic nature of this class of transformations to hold is restriction of numbers of molecules in space and time of processes of reproduction.

It is satisfactory that an operation under this rule must be a cyclic process which involves the action of large molecules and is closed (does not exchange mass and energy with the environment). Schrödinger (1945) has emphasized the importance of molecular stability and negative entropy (or information conservation) in genetics. Expressed in another way, the quantum rule says that "like will produce like cyclically" or that genetic information is conserved cyclically.

Logarithmic spiral arrangements in sunflower and pine cone phyllotaxis, in marine crustacean shells, in mammalian cochlear and labyrinthine apparati, in neurotome and myotome development, in human and many other animal body proportions (golden section), etc. suggest a correspondence with the Fibonacci series of integers which approximate the combined logarithmic and linear intervals of some epigenetic transformations. The logarithmic aspect of the intervals in epigenetic transformations in more complex organisms permits a fine correspondence for the requirement that capabilities have to be traded off or given up to another bion or cell for augmentation of special capabilities in a particular bion or cell. The bion or cell still must be able to reproduce itself, and the cyclic nature of the class of transformations still allows for the system to duplicate itself. The nature of this particular class of transformations fits amazingly well the form of quantum rules for processes of epigenesis.

The discontinuous or multivalued behavior of nerve and muscle membrane depolarization, for example, requires open metabolic mass and energy transport with the environment. This has been deduced as the form that one class of quantum rules has to take in the quantum mechanics which involves processes with larger indeterminacy. The form of this class of quantum rules is: the class of metabolic mass and energy transport processes obey a class of transformations with an asymmetric (multivalued) nature. The phenomenological condition for the asymmetric nature of this class of transformations to hold is radical openness of mass and energy transport with environment at ordinary temperatures and pressures.

SUMMARY OF THE FORMS OF THE CLASSES OF TRANSFORMATIONS AS QUANTUM RULES

Although at this stage it is somewhat premature to make this series of generalizations, it will be important to search for the authenticity of the following set of correspondences:

a) Genetic information transport processes follow single-valued transformations which reflect the cyclic nature of the processes, undergo closure, and are identical or one-to-one with addition of linear intervals. A paraphrase of the form of this class of quantum rules is that "genetic information in processes of living units is conserved cyclically in time and space."

b) Epigenetic information transport processes follow single-valued transformations which reflect the cyclic nature of the processes, undergo closure, and are one-to-one with addition of logarithmic intervals (or combined logarithmic and linear intervals). A paraphrase of the form of this class of quantum rules is that "epigenetic information in processes of living units is expressed cyclically in time and space."

c) Metabolic mass and energy transport processes follow non-single-valued transformations which reflect the asymmetric (irreversible, non-equilibrium, or open) nature of the processes and which undergo closure. A paraphrase of the form of this class of quantum rules is that "mass and energy in processes of living units is exchanged asymmetrically with environment."

ENVIRONMENTAL PARAMETER SELECTION AND EPIGENESIS

It is important to proceed further with the point that the above classes of genetic, epigenetic, and metabolic transformations have parameters or "subclasses." (See examples XI and XII on page 72.) This permits a distinction between (*a*) *environmental parameter selection* and (*b*) *epigenesis* which has been a source of confusion to biochemists and biologists.

External interactions of the environment with living units or bions may be put into correspondence with environmental parameter selections, that is, selections of subclasses of the major classes (genetic, epigenetic, metabolic) of bionic transformations.

Epigenesis may be put into correspondence with that major class of transformations which has been reserved here for internal information transport beyond the genetic variety that arises in giving up functions to augment others among bions in the formation of complex cells and tissues in organisms. This does not come into existence until the appearance of complex biological organisms (see Chapter 6).

These comments will be useful as background for the discussion of "bions" in the next chapter, where a more specific distinction is made concerning "bions" and cells.

GENES

The term, "gene," is used in the present writing in much its old or classic sense, as the hypothetical unit of inheritance. The "genome" of a cell is the totality of identifiable characteristics, overt or potential, which are passed between generations of cells.

Each gene, at its basic level of characterization, is represented by an element (operand) of the genetic transformation, as referred to in this manuscript. The expression of a gene can be regarded as the selection of a subset from the genome by means of parameter selection.

HORMONES

The multiple nature of the influence of hormones (Stetten, 1959, and Davidson, 1965) strongly suggests an indeterminate molecular basis of operation and seems to be equivalent to making parameter selections among the genetic, epigenetic, and metabolic classes of bionic transformations. Although hormones are apparently part of an interbionic communication process for parameter selection which has an indeterminate molecular basis, all parameter selection is not restricted to hormonal action.

References

Ashby, W. R. 1936. *An Introduction to Cybernetics*. New York: Wiley.
Davidson, E. H. 1965. *Scientific American* **212**: 36.
Mayer, E. 1961. *Science* **134**: 150.
Schrödinger, E. 1945. *What is Life?* New York: Macmillan.
Stetten, D., Jr. 1959. In *Biophysical Science*, ed. J. L. Oncley. New York: Wiley.

PART II

IMPLICATIONS OF UNSUSPECTED QUANTIZATION IN BIOLOGY

CHAPTER 5

THE BION

Basic atomistic living unit

The proposition must be confronted as to whether there exist atomistic units in biology which combine at least two classes of processes, reproduction and metabolism. At the present time, having only meager knowledge of the size of the $\in$ factor of indeterminacy relations to guide us through "time-energies" or "space-momenta" of fundamental emergent processes of these units, we must ask the question, "Are there living units that exist under conditions of sufficiently few events in space and time, and sufficiently open transport of mass and energy to and from environment at ordinary temperatures and pressures, that quantum description is required?" An affirmative answer can be plausibly made in view of the arguments of Chapter 3 and by virtue of alignment of the above criteria with certain freely or singly living, microscopic, biological forms (blue green algae, nonnucleated bacteria, and *Mycoplasmatales*) and with some subcellular organelles in organized cells. The latter include nuclei, which are mainly concerned with genetic information conservation; plastids (chloroplasts and leucoplasts) and mitochondria (Gibor and Granich, 1964, and Granich, 1965), which are primarily concerned with metabolic mass and energy conversion; centrioles and kinetosomes, which seem to have principally motor and sensory functions; and possibly microtubules (Porter and Tilney, 1965), which seem to be elastic cytoskeletal elements with some connection with intracellular movement.[1] The name *bion* will be given to the basic atomistic living unit which has the minimal capabilities of reproduction and metabolism, tentatively possesses both nucleic acid and protein molecules, and is limited in size to about 0.1–3.0 microns (some dimensions may be greater than this). The coupling of bions into one unit or cell together with specialization in component bions, in the manner of "molecularity" of atoms, allows for great versatility in structure and function of cells and cell aggregates as living organisms. This basic unit of biology is not unlike the atom of physics and chemistry. It, however, is associated with a broadened application of quantum theory.[2]

With the general biological term, cell, and the term for subcellular particles, organelle, already in use, it might appear redundant to attempt to coin a new term, bion. The justification is that it is necessary as a unifying measure, since all cells are not classifiable as bions and all organelles are not classifiable as bions. For instance, *Mycoplasmatales*, blue-green algae, and nonnucleated bacteria are singly living bions or unibionic organisms. Paramecia, amebae, and Euglena are examples of singly living, organized, unicellular organisms which are not bions, but are made up of bions. Nuclei, plastids, kinetosomes, centrioles, and microtubules are examples of subcellular organelles which are bions. Endoplasmic reticulum and the Golgi complex are subcellular organelles (Fawcett, 1966) which are not, or very likely are not, bions.

A bion is necessarily limited in size but is not necessarily limited to the size of a cell or an organelle. Confusion is encountered in any cellular hypothesis of a bion merely because of numerous instances of very large cells: such as the giant cells of *Acetabularia*, *Valonia* (algal) cells, fertilized frog eggs, giant cells of mammals, *Porospora gigantea*, etc. The same is true in any organelle hypothesis of a bion in view of the very small size of microsomes and very large size of some lysosomes.

[1] Still other particles, at present ill-defined as to "microphysiology" and "microgenetics" and occurring in specialized cells, may tentatively be classified as bions pending further investigation. A partial list may include the kappa particles of paramecia, the cortex of the fertilized amphibian egg, and the nucleolus.

[2] After the completion of this book, we learned that Wilhelm Reich had employed the term "bion" to name "fundamental living units" (*Cancer Biopathy*. Rangley, Maine: The Wilhelm Reich Institute. 1948). In our work, the theoretical basis for the existence of the bion, and the operational meaning of the term, are quite different from that used by Reich, however. (As long as the operational meaning of "bion" is retained as described in this chapter, another word such as viton, vion, bio-atom, biological atom, small living particle, or the like could be substituted to designate this fundamental living unit.)

Among the independently living forms, some are quite small, such as the *Mycoplasmatales* type of bion. The *Mycoplasmatales* are always very small (i.e. 0.1 micron). They may become bigger in one dimension in some instances, but this dimension does not exceed several microns. They have no cell wall; they do have a plasma membrane. There is scarcely any distinguishable structure inside them. For instance, there is no other envelope or membrane inside the plasma membrane. Blue-green algae or *Myxophyceae* are also small. They have no nuclei. They have layers of chlorophyll, but these are diffusely distributed. The *Schizomycetes* or fission fungi or bacteria have no or indefinite nuclei and do not undergo mitosis. Most are less than one micron in size.

Among the unibionic organisms a large multiplicity of tropisms or directed motions have developed. Also, bions have very versatile synthetic capabilities. In dilute media, they sometimes have quite heterogeneous capabilities for both synthesis and motion, which has the same effect as providing a greater effective volume.

If a bion warps toward special capabilities, it loses others. Expression of a capability such as motion may be associated with variation in form from monotrichous flagellation to multitrichous flagellation. In such bions locomotive capability is traded off against loss of metabolic capability. With its augmented movement, a bion with multiple flagella may lose a certain vitamin-synthesizing capability, but has the reciprocity of reaching or running into this vitamin supply exogenously. There is also a profound implication with respect to sexual fertility and sterility where motility is contributed by the male and metabolic power and control are contributed by the female.

It is a short step from reciprocating or trading off capabilities between a colonial or singly living bion and its general environment in order to augment or specialize bionic function, to losing or trading off capabilities in cooperation among bions in order for the bions to augment or specialize function in such a manner as to form a cell. This second type of trading or reciprocity requires that one bion maintain the organizing, long-term information principle for the cell. At least one bion (the nucleus) must retain the capability for reproduction of the organized cell. However, every bion which joins with one or more other bions within an organized cell must have the ability to reproduce itself within the interior of the organized cell. In this case, the inside of the cell is now the environment of the bions. To reiterate, the combined emergent capabilities of reproduction and metabolism are still present in a bion, even if it is inside a cell as a nucleus or mitochondrion or kinetosome or centriole or plastid or microtubule.

In this second step of trading off information to environmental bions in order to augment capabilities, the changes may be vast. For instance, some capability such as motion may almost completely occupy a bion (say a bion with flagella). This capability of motion may be greatly augmented, while oxidative phosphorylation is practically given up to another bion which relinquishes control of its own motion by gaining great oxidative-phosphorylating power. Even in this case, the original definition of a bion is not lost, for the bions must still have capabilities for reproduction and metabolism. This is maintained despite the possibility in a multibionic cell in a multicellular organism that all other functions have been traded off for augmentation until the bion is virtually unifunctional. The difficulty or delay in reproduction because of the previous reciprocity may be great, but reproduction is possible under appropriate conditions.

COUPLING OF BIONS IN HIGHER ORGANISMS

The point was established in Part I that there is indeterminacy of motion of molecules in many living processes, with discrete behavior emerging. Transformations may be employed to represent these quantized or emergent processes. Cybernetic transformations infer a machine-like input-output manner of description. When using transformations which are similar to those of cybernetics for representation of *emergent* living processes, the distinction must be very carefully maintained that the processes being represented are only machine-like, since these living processes have an indeterminate basis of operation.

This makes possible a far more precise discussion of the ways in which bions may interact in the formation of higher organisms. One way is in the form of *selection of parameters* or subclasses of the classes of transformations which represent the emergent living processes in a bion. This occurs through interaction of a bion with the general external environment (which may include both extracellularly derived agents and those from other parts of a cell which contains a bion). A second way is in the form of what we term an *epigenetic* class of internal transformations within the bions which represent additional emergent processes that appear with formation of multibionic organized cells in multicellular tissues of colonial organisms, quasi-organismic organisms, or fully organismic organisms. Parameter selection also occurs with respect to the epigenetic class of emergent bionic processes, as it does with the genetic and metabolic classes of emergent processes.

Epigenesis has a more or less classical meaning in biology, namely, environmental interaction with zygote in order to complete the information which is necessary for development of the zygote into an organism. This is contrasted with the preformation theory, which assumes a completely preformed "germ" that is so small as to be invisible inside the zygote, but which enlarges to visible size in development. An evolved version of the preformation theory is the present day concept of the gene with the complete information for form and function of an organism which is stored in genetic arrays of chromosomes that are held under gene repression in the zygote ready for reading out under derepression during development. Elsasser (1961) attempted to make a distinction concerning emergent living processes, which he called *biotonic*. Later (1962), at Commoner's suggestion, he termed them epigenetic. There is a risk of great confusion unless the following comments are introduced at this stage. Our "non-Gaussian" emergent living processes (which include processes of reproduction, metabolism, and epigenesis) are consistent with Elsasser's biotonic processes, as he originally described them. We use the term, epigenetic, specifically to name the emergent living processes of bionic specialization and augmentation of function which are achieved by giving up or losing of information among interacting bions in multibionic cells on an indeterminate basis of operation (i.e., under conditions in which molecular motion is indeterminate).

There is a third way of interaction of bions which contributes to the organization of more complex organisms. Bions may be considered as unitary "black boxes" which are holistically capable of determinate inputs and outputs. These determinate inputs and outputs include chemical concentrations, osmotic gradients, long-range forces (see Chapter 7), bionic radiation (see Chapter 8), etc.

Not all indeterminate processes involving molecular motion are restricted to bions. (Analogously, not all indeterminate processes involving motion of elementary particles are limited to atoms.) Indeed, the requirement for quantized description of certain reproductive and metabolic processes was shown without reference to the bion as a unit. Therefore, the disclaimer must be introduced that, should the bionic concept not prove valid in generalization or useful in application, many biological processes with their peculiar phenomenological conditions still require a quantal mode of description. The point to be made is that indeterminacy of molecular motion may extend beyond processes occurring in the interior of a single bion. Spatial (volume) and temporal limitations may occur in interconnections between units, forming a hierarchy. This is to say that in the "doubling back" between bions to form cells and between cells to form tissues or organs, there may be modes of dynamic coupling which involve indeterminacy of molecular motion. *Retroversion* is the term that we have given to this incorporation of indeterminate molecular processes into major modes of interaction between biological subassemblies. Such retroversion leads to emergent behavior of the cell or organ itself. Here, as within bions, determinate factors, such as surface tension, diffusion rates, and so forth are very important, but the distinguishing feature is that a fourth way of interaction may contribute to hierarchical organization in the form of a new class of emergent processes. By virtue of retroversion, capabilities emerge for directed motion, secretion, neural synaptic transmission, psychic states (see Chapter 9), etc.

In the above four ways of interaction, bions as "indeterminate machines" join in complex organismic arrangements *nondestructively*, that is, with stability of the bions.

A coherent classification of the full spectrum of living organisms can be based on these four ways of interaction (Chapter 6). Before this, however, a consideration of the limits of bionic size must be presented.

INDICATIONS OF LIMITS OF SIZE OF BIONS

There is an upper limit to the amount of information that *can* be handled on a mechanistic basis in a limited space in a limited interval of time (limit of number of microparticulate oscillators oscillating in a space in a limited time). There is an upper limit of the number of particles in a space and time which is still consistent with information conservation rules after the manner of non-Gaussian information handling. (This is equivalent to saying that the number of energy states in the statistical ensemble must be kept radically small during assemblage of the ensemble.) An upper limit to the size of bions would be expected to correspond with the build-up of a sufficient number of particles in a limited time and space to give sufficient number of events to be Gaussian, if it is recalled that the non-Gaussian condition is required for emergence of genetic and epigenetic information handling processes which can be described by closed cyclic transformations.

There is also a lower limit to the amount of information that *must* be handled by particles in a limited space in a limited interval of time which is consistent with rules of sufficient asymmetry of metabolism, that is, after the manner of sufficiently open interchange of energy and mass with the environment to constitute non-Gaussian metabolic turnover. (This is equivalent to saying that the number of energy states available for occupancy in the statistical ensembles must be radically large, necessitating a certain minimum space.) A lower limit to the size of bions would be expected to correspond with the loss of sufficient openness of mass and energy transport with the environment in a limited interval of space and time to be Gaussian, if it is recalled that the non-Gaussian condition is required for emergence of that nature of metabolic processes which can be described by transformations that reflect the radical asymmetry.

There are a number of indications as to what these upper and lower limits of bionic size actually are:

1. The interactions between long-range forces and thermal collisions, as taken up in Chapter 7, are projected as setting limits of bion size between approximately a thousand angstroms and several microns. Too large a bion would favor the forces of random thermal collisions with a bion. (This would tend to overcome individual intrabionic coupling forces which tend to become attenuated as a bion grows larger.) Also, if a bion is too large it is not functional, for sufficient amounts and varieties of electromagnetic input and output in the range of visible light or ultraviolet frequencies do not accumulate for the holistic operation of that bion. Existence of too small a bion does not constitute a sufficient condition for the increase of transitional probabilities and of macromolecular coupling energies which is necessary for formation of a bion. Also, if a bion is too small, it is not functional in that sufficient amounts and varieties of electromagnetic input and output in the range of thermal or infrared frequencies do not tend to accumulate for holistic operation of a given bion.

2. Limitation of the size of bions (for holistic or quantal processes) is connected with the considerations of simultaneity and causality. The most rapid transmission of action known is that of electromagnetic radiation (at a speed of roughly 3×10^{10} centimeters per second). The electromagnetic frequencies of virtual photons which are exchanged between oscillators in molecular aggregates as bions under non-Gaussian conditions are projected as ranging between the approximate 10^{14} vibrations per second of hydrogenic stretching frequencies (befitting the bionic structure which is necessary for catalysis and synthesis), and the approximate 3×10^{15} vibrations per second of predissociation and dissociation frequencies (befitting the smaller molecular turnover of bionic metabolism). Several thousand angstroms to several microns is the range of the longest distance which can be transversed within one such vibration time by a photon at the maximum possible speed. This is the limit for holistic operation on an indeterministic basis at ordinary temperatures. Beyond this general limit of sizes, external knowledge of simultaneity and causality with respect to vibratory phenomena within the bion (interphenomena) is restricted or lost. Likewise, internal knowledge of simultaneity and causality with respect to vibratory phenomena outside the bion is restricted or lost. (These, by the way, are excellent conditions for "two quantum processes" (see Goppert-Mayer, 1931) permitting interpretation of photosynthesis and chemiluminescence in biology on simpler terms. This offers an approach to explication of the baffling Emerson effect in photosynthesis, the Pasteur and Crabtree effects in intermediary metabolism, and the peculiar peroxide requirements in chemiluminescence.)

3. Simple living forms (blue-green algae, nonnucleated bacteria such as *E. coli*, and forms without cell walls such as *Mycoplasmatales*) and cellular inclusions (nuclei, centrioles, kinetosomes, plastids, mitochondria, and microtubules), which we have called bions for biological reasons (capability of both reproduction and metabolism), have sizes that range between several thousand angstroms and several microns. This is to say that there is an actual biological correspondence with the sizes of bions as theoretically projected.

4. When we utilize the light microscope and some type of minimally injurious system for dissipation of heat released to the biological environment by the microscopic viewing system, the smallest sized units of bio-

logical structure still visible in the relatively viable state fall into the range of several microns to several thousands of angstroms.

References

Ashby, W. R. 1963. *An Introduction to Cybernetics.* New York: Wiley.
Brillouin, L. 1962. *Science and Information Theory.* New York: Academic Press.
Elsasser, W. M. 1961. *J. Theoret. Biol.* **1**: 27.
———. 1962. *J. Theoret. Biol.* **3**: 164.
Fawcett, D. W. 1966. *The Cell.* Philadelphia: Saunders.
Gibor, A., and Granich, S. 1964. *Science* **145**: 890.
Goppert-Mayer, M. 1931. *Ann. d. Physik* **9**: 273.
Granich, S. 1965. Report of 1964 Symposium, Am. Soc. of Naturalists (at Montreal Meeting of Am. Assoc. for the Adv. of Sc.). *Science* **147**: 911.
Porter, K. R., and Tilney, L. S. 1965. *Science* **150**: 382.
Schrödinger, E. 1945. *What is Life?* New York: Macmillan.

CHAPTER 6

APPLICATION OF THE BIONIC CONCEPT

The bion as an atomistic unit: application in a comprehensive classification of living organisms and in collation of disconnected biological phenomena

An application will now be made of the concept of the bion as an "atomistic" unit of biological organization. The feature of "atomicity" of the bion can be shown to be capable of serving as a basis for a *comprehensive* classification of living organisms. The bionic concept can be utilized for collation of hitherto seemingly disconnected biological phenomena. These include: the fundamental difference in development of plants and animals; a basis for nuclei in biological cells; a foundation for sexual conjugation of germ cells; an understanding of "ontogeny recapitulating phylogeny" in animal embryogenesis; and an inherent distinction between genetic and epigenetic information handling. Others are discussed in Chapter 9. The bionic concept can be employed for prediction. Chapter 5 on the bion and Chapters 7, 8, and 9 deal with various predictions arising from the application of quantum theory and the bionic concept in biology.

THE BION AS AN ATOMISTIC FEATURE IN COMPREHENSIVE CLASSIFICATION OF LIVING ORGANISMS

I. Unibionic Organisms

Unibionic organisms encompass blue-green algae (Myxophyceae), fission fungi or bacteria having no nuclei (Schizomycetes), including the pleuropneumonia-like organisms (Mycoplasmatales). Free-living bions are single-envelope systems, have no nucleus, grow on cell-free media, have at least capabilities of reproduction and metabolism, are limited in size to approximately 0.1–3.0 microns (in at least one dimension), and possess large molecules (tentatively believed to be at least nucleic acids and proteins). Viruses and rickettsia do not seem to be free-living bions. Rickettsiae are of bionic size, have a single envelope with no nucleus, but require a cellular environment for growth. Viruses may be about the size of bions, but possess no plasma membrane. Even for replication ability, they depend on the environment of cellular hosts or artifically supported metabolism. Viruses are analogous to subatomic particles, and rickettsiae to chemical free radicles trapped in a matrix. We suggest that viruses may be considered as sub-bionic particles and rickettsiae as "parasitic" bions, entrapped in cells.

Colonial unibionic organisms may occur. *Myxobacteriales* and some of the *Thiobacteriales*, *Spirochaetales*, *Cyanophyceae*, and iron bacteria are examples.

II. Unicellular Organisms

Multibionic cells are of two principle types, primitive and organized. Each type may exist as singly living or colonial organisms. They are at least two-envelope systems. They are capable of division by mitosis and meiosis, as well as amitosis. This step of multibionic living is analogous to "molecularity" in chemistry.

A. Primitive Unicellular Organisms Primitive cells possess only a nucleus and cytoplasm (and possibly a few chloroplasts or mitochondria).

1. *Singly living primitive unicellular organisms* include the green algae (Chlorophyceae), the ascus fungi (Ascomycetes), some protozoans (Sporozoa), and karyotic fission fungi (Schizomycetes).

2. *Colonial primitive unicellular organisms* include the coenocytic green algae (*Vaucheria*), coenocytic algae-like fungi (Phycomycetes), and coenocytic slime molds (Myxomycetes) with a swarm stage.

B. Organized Unicellular Organisms Organized cells may exist as singly living or colonial organisms. They have nuclei and many different bions. Organized cells have the ability to aggregate certain functions into each of the variously occurring bions. These bions or subcellular units still

Table 3. Comprehensive classification of living organisms based on the bion

(Sub-bionic particles—viruses, bacteriophages)
(Rickettsia—"parasitic" bions)

- I. Unibionic Organisms
  - Blue-green algae (Mycophyceae)
  - Akaryotic fission fungi (Schizomycetes)
  - Mycoplasmatales
  - } Includes part of Thallophyta (Vegetable Subkingdom)
- II. Unicellular Organisms
  - A. Primitive Unicellular Organisms
    (Parameter selection, reversible; 2 or a few bions; nucleus and cytoplasm, possibly chloroplasts)
    - 1. Singly Living Primitive Unicellular Organisms
      - Green algae (Chlorophyceae)
      - Sac-bearing fungi (Ascomycetes, yeasts, Fungi Imperfecti, some molds)
      - Some protozoa (Sporozoa)
      - Karyotic fission fungi (Schizomycetes)
    - 2. Colonial Primitive Unicellular Organisms
      - Coenocytic green algae (*Vaucheria*)
      - Algal fungi (Phycomycetes)
      - Coenocytic slime molds (Myxogastrales or Myxomycetes of subclass Mycetozoa)
  - B. Organized Unicellular Organisms
    (Parameter selections from environment which are usually reversible; many bions)
    - 1. Singly Living Organized Unicellular Organisms (Protozoa)
      - Flagellates (Mastigophora)
      - Ciliates (Ciliata)
      - (Sarcodina) and (Suctoria)
    - 2. Colonial Organized Unicellular Organisms or Reversible Multicellular Tissue Organisms
      - *Volvox* (Volvocidae), Mesozoa
- III. Multicellular Tissue Organisms
  - A. Quasi-Organismic Organisms
    (Beginning of epigenesis, with some specialization and only a little internal degeneration of bions; epigenesis relatively late, i. e. after or concurrent with setting aside of gamete forming and embryonic germ tissues for reproduction of new organisms; mostly nonspecialized or *primitive tissue* with development of only a few organs)
    - Red algae (Rhodophyceae)
    - Brown algae (Phaephyceae)
    - Club algae, rusts, smuts, fleshy and woody fungi (Basidiomycetes)
    - } Includes part of Thallophyta (Vegetable Subkingdom)
    - Higher plants
      - Bryophyta (Mosses and liverworts)
      - Pteridophyta (Ferns and related plants)
      - Spermatophyta (Seed-bearing plants)
      - } Embryophyta (Vegetable Subkingdom)
    - Cellular slime molds (Acrasiales)
  - B. Fully-Organismic Organisms
    (Definite epigenesis and *specialized tissue* formation, with epigenetic change early, i.e. before sexual maturation of the organism; multiple germ layers differentiate in embryo; ontogeny recapitulates phylogeny; specialized tissues develop into many organs in a number of systems)
    - Sponges (Porifera)
    - Invertebrates
    - Vertebrates

possess both reproductive and metabolic capability, even though parameter selections modify bions and bions perform specialized functions within organized cells.

1. *Singly living organized unicellular organisms* include the protozoans with external motion, Mastigophora (with a few long flagella of kinetosomal origin), Ciliata (with many short ciliae of kinetosomal origin), and Sarcodina (with pseudopods in which the cortex of the cell or centriole-like bodies produce movement of a large amount of cytoplasm).

2. *Colonial organized unicellular organisms* include, for example, the colonial flagellates (Volvocidae) of the Mastigophora class, such as the *Volvox*, which have complete reproductive and metabolic capabilities in each of their some 20,000 cells. Form and function depend on parameter selection which stems from position in the colony. The cells, in other words, have undergone no epigenetic change and assume somatic or sex function for the organism dependent on physical position in the colony and parameter selection accruing therefrom. Any cell may be reversibly separated and reproduce the organism or metabolize independently. These may also be called reversible multicellular tissue organisms.

III. Multicellular Tissue Organisms

Organized cells can live together not only in reversible colonies as multicellular tissue organisms, but in multicellular tissues of quasi-organismic organisms and in multicellular tissues of fully organismic organisms. This step of multicellular living is analogous to "macromolecularity" in chemistry.

In fully organismic organisms, the classes of processes of organized cells are dependent on one another. Having undergone irreversible epigenetic change, the bion may be said to "breed true," within the organized cell, propagating a tissue line upon cell division.

A. In the tissues of quasi-organismic organisms, parameter selection prevails, as it does in colonial tissue living. Most cells remain pluripotent and nonspecialized, and could, under appropriate conditions, reproduce the whole organism (Halperin, 1965). The size of most cells has not been restricted under epigenetic change. Size, shape, and function of these cells is dependent upon parameter selection in any particular situation. However, some epigenetic divisions of bions may occur. These epigenetic divisions which lead to specialized multibionic cells, primitive multicellular tissues, and a few permanent organs occur relatively late, or concurrently and cyclically with development of germ cells and reproductive tissues. These epigenetic divisions occur in a number of widely scattered meristemic tissues in plants. At these sites, some pluripotent, nonspecialized cells are maintained within the growing meristem in order that information is conserved for reproduction of a whole new organism. This may be in the form of egg cells or buds. These epigenetic divisions are seasonally or environmentally recurrent. There is no development, or only primitive development, of embryonic tissue layers.

Quasi-organismic organisms include red algae (Rhodophyceae), brown algae (Phaephyceae), club algae, rusts, smuts, and fleshy and woody fungi (Basidiomycetes), cellular slime molds Acrasiales), and all higher plants (Embryophyta).

B. In the tissues of fully organismic organisms, in addition to parameter selection, many epigenetic divisions of bions occur relatively early in the life history of the organism (before sexual maturation of the adult form). These are exhibited as the zygote divides into tissue germ layers in the embryo with subsequent development of many fully differentiated organs in a number of systems. The epigenetic divisions may be staged cyclically in embryogenesis, as ontogeny recapitulating phylogeny, and at puberty. Larval formation, metamorphosis, and growth spurts are related to parameter selection. Some cells are set aside in the pluripotent, nonspecialized condition in which resides all the necessary information for reproduction of a new organism. Some of these cells develop into egg and sperm cells (gametes) after most epigenetic divisions have occurred in the organism. Some of these cells may undergo delayed epigenetic development into supportive reproductive tissues for the gametocytes, the fertilization process, and development of the new embryo.

Fully organismic organisms include sponges (Porifera) and all metazoan animals.

Figure 7 traces the development of fully organismic metazoan animals in contrast with Figure 8 which traces the development of quasi-organismic *Embryophyta* plants.

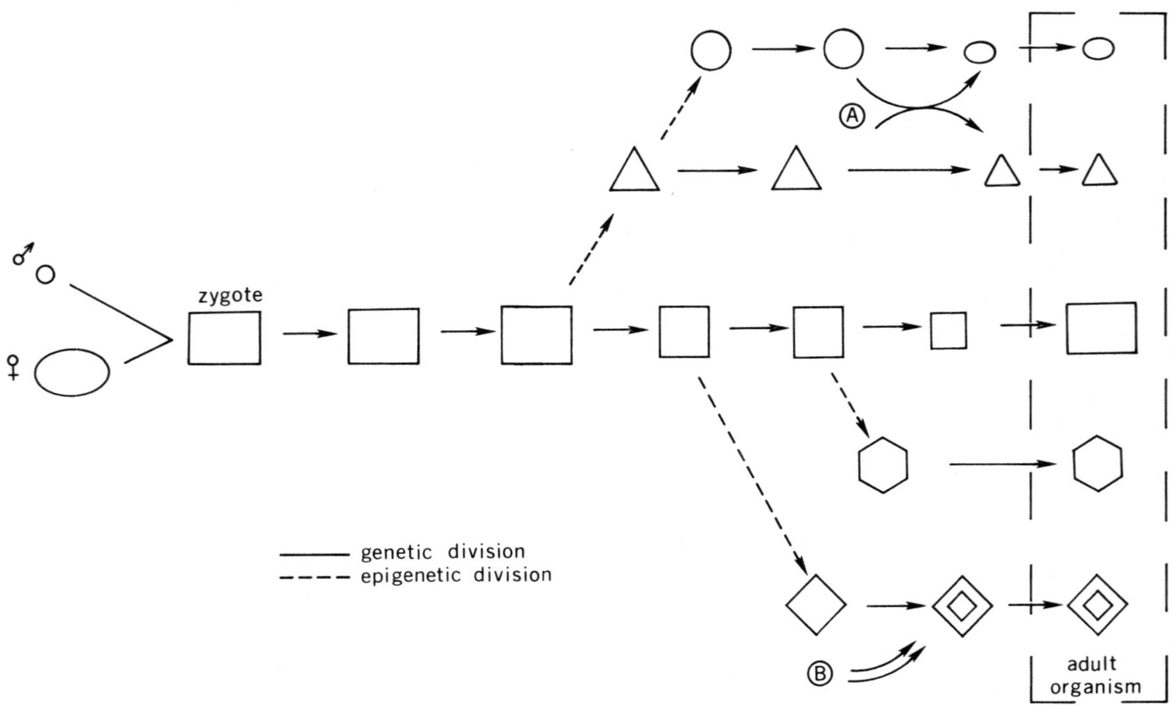

Figure 7. Development of fully-organismic organisms (*animals*). *A. Interactions occurring between cells. B. Environmental interactions* (*contact and distant*).

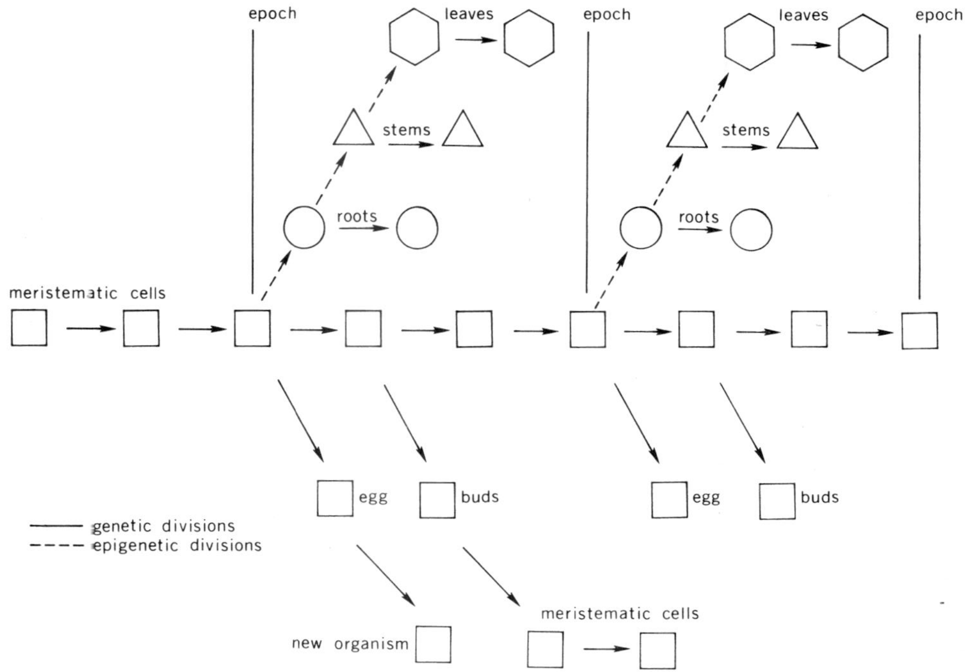

Figure 8. Development of quasi-organismic organisms (*plants*).

DEVELOPMENT OF FULLY-ORGANISMIC METAZOAN ANIMALS

In metazoan animals, female and male germ cells undergo sexual conjugation with exchange of bionic material, particularly of nuclear bions, followed by meiotic reduction division of nuclear chromatin material of the fertilized egg or zygote.

The zygote represents an entirely new counting of the number of occupied energy states in bions of a developing organism. Two major phenomena occur that might be overlooked: *maternal inheritance* (probably from division of bions other than the egg nucleus) and information exchange leading to *heterotopy*. (The antithesis to the latter is the loss of hybrid vigor in plants that are maintained asexually.) *Reduction division* of the zygote is followed by "genetic" divisions of new cells which occur in the blastulation and gastrulation stages. After these very early stages of embryogenesis, *epigenetic divisions* as well as *genetic divisions* begin to occur. (Environmental conditions are important to both, also. This means that, through various epigenetic divisions, pluripotent or primitive cells begin to specialize as stem cell types. After epigenetic division, the specialized new cell lines "breed true" (reversibility to former type now being impossible under ordinary conditions) as the germ tissue layers of the embryo. Here, there is trading off or loss of information and augmentation of function among bions in cells, with cells being limited in volume and joining as specialized tissues in fully developed organs. The environmental conditions under which adult cells and tissues form are of two general types: *organismic* and *nonorganismic* or *independent*. The organismic environmental conditions under which bions and cells exchange information are summarizable in two main types of relations: *contact relations* and *distant relations*. Contact relations include those of shape and pressure, adhesion, electrical interaction, molecule or template, and long-range forces. The distant relations include blood and lymph circulatory, neural, and hormonal. The independent or "world" or nonorganismic influences are thermal, photic, gravitational, etc. Coupling of bions and cells occurs mainly through parameter selection and epigenesis. Genetic divisions continue to occur. In some primitive, pluripotent cells genetic divisions occur without epigenesis. In this manner some pluripotent cells are retained in the developing embryo and on into the adult organism, dividing only genetically. The adult metazoan is then composed of variously differentiated cell types, lacking zygotic or genetic information, and of some nondifferentiated cell types with all the zygotic information. Some of the nondifferentiated or pluripotent cells become egg or sperm cells. During the reproduction process these cells are under the influence of the remainder of the differentiated cells and tissues of the adult organism and their environmental conditions. Some pluripotent cells are involved in the very special development of those "reproductive" tissues which are required in support of the germ cells and for holism with the rest of the organism during the reproductive processes.

DEVELOPMENT OF QUASI-ORGANISMIC PLANTS

A basic difference between plant and animal organisms is in the when (temporal relation) and where (spatial localization) of the pluripotent cells. In animals, they are in one place usually, localized early in the life of the organism, and are not functional in the reproduction of the organism until after its epigenetic maturity. Most of the animal organism's cells undergo epigenetic change very early with embryonic differentiation. In plants, the pluripotent cells are at many places in the organism, i.e. at the growing fronts as meristematic tissues where they undergo development into various kinds of buds. The buds may develop into flowers with ovary, ovules, egg cell, anther, pollen sacs, and pollen mother cells, with reduction division of egg and pollen cells and pollination. Sexual conjugation or fertilization may then occur followed by zygote division without reduction in 2 N chromosome numbers, embryo formation in a seed, and fruiting. Embryogenesis is of a limited type as seeds in the higher plants. (The buds can also form new organisms if placed in an appropriate environment, or can be transplanted to other organisms.) Mainly, however, the buds form new apical meristems before or concommittant with limited epigenetic division of some of their cells for specialization into what are called permanent tissues of the budding organism. The genetic and epigenetic divisions wind up in the formulation of still rather primitive or only partly differentiated tissues in a few organ types: roots or "hold fasts," stems, leaves, and flowers. These genetic and epigenetic divisions occur cyclically, often under the influence of seasonal rhythms of temperature, substrate availability, light, gravitational force, etc. To reiterate for clarity, the meristems are groups of primitive or pluripotent cells with all the information necessary for starting the development of another organism. The bulk of these primitive cells divide only genetically and do not form new organisms. Together with lesser numbers of epigenetically specialized cells, they form permanent tissues or cell aggregates. The cells of the permanent tissues then divide only genetically and seem to live in a colonial or quasi-organismic type of arrangement.

THE BION AND COLLATION OF LIVING PHENOMENA

The disjunction in time and space between epigenetic development of somatic tissues and genetic development of egg and fertilizing germ cells (gametes) and mature reproductive tissues is seen to be a *fundamental difference between plants and animals.* The animals tend to undergo early, massive, and extensive "somatic tissue" epigenesis and later, less massive, more localized gamete and reproductive tissue maturation. The plants tend to undergo early, massive, and extensive gamete and reproductive tissue maturation and later, much less massive, more localized epigenesis in the development of permanent tissues.

The very early, massive, extensive, and cyclic epigenetic divisions of cells in animals is exhibited as the development of definite germ layers, embryonic tissues, and organ systems which is acted out as *ontogeny recapitulating phylogeny in embryogenesis*, as a series of epigenetic cell divisions occur along with genetic divisions.

The relatively late, much less extensive and less massive epigenetic divisions of cells in plants is exhibited in several ways: first, in the ability of any cell with appropriate metabolic support to reproduce the whole plant organism (Halperin, 1965); second, in the more colonial-like living and development of only a few organs; third, in the pleomorphism of the cells; and fourth, in the more primitive, less specialized embryogenesis.

A basis for nuclei in cells is to be found in intracellular bions which conserve information in the reproduction of the cell as a unit. Each of the constituent bions of a cell is capable of division and, indeed, must divide within a characteristic time in order to restrict the number of occupied energy states to achieve molecular indeterminacy. The nucleus, moreover, seems to be the bion concerned with the information of the cell as a unit or a set of coupled bions. The role or work function of the nucleus is seen to be: *integration* of the work function of other bions into the economy of the whole cell; *synchronization* of cell division with bionic division; and efficient *transfer* of information to the daughter cells upon division.

Despite the quantized discrete nature of information transfer from cell to cell, chance molecular aggregations may tend to persist over many asexual generations. Such aggregations can be seen to contribute to a possible information loss or degradation. *Conjugation* with the resulting dynamic interaction of similar but slightly different molecular species (and concentrations) is an effective method for countering such loss of biotonic information. *Sexual conjugation* is a further elaboration to insure this molecular mixing in the zygote or beginning organism. Thus, the multicellular organism insures that conjugation occurs at some point of its life cycle since the germ plasm may have undergone repeated division within the organism itself.

The seeming anomaly of *amitotic division* of multibionic cells deserves comment. Amitotic division can act as one form of information conservation process for a cell. It may be more prevalent than appreciated, since most observation of cells is under fixation, staining, and microscopic manipulative technics which tend to have an element of pragmatic selectivity which favors viewing the cells during mitotic or meiotic division. Amitotic division may occur in specialized or epigenetically transformed cells under the subtle condition that the reproductive and metabolic capabilities of a cell have been warped toward placing the kinetsome to the outside for ciliary action (bronchiolar epithelium, etc.) or into sarcomeric distribution for striated muscle action, for instance. The centriole function in motion of nuclear bionic chromatin matter in mitosis may thus be lost. It is not surprising to observe that any regeneration of striated muscle cells or bronchiolar epithelial cells is either abortive or amitotic (Maximow, 1930). Attention is called to the laborious studies of amitotic cell division in the ovaries of sheep tapeworm proglottids by Child (1907).

The distinction between genetic and epigenetic information transport is inherent in the restriction of number of occupied energy states of molecular motion which necessitates an indeterminate basis of operation of bionic processes. Under this proscription of molecular motion, conservation of genetic information occurs cyclically in the form of emergent processes of reproduction of an organism, and loss of information can occur among bions of an organism which is composed of interdependent cells, permitting augmentation of bionic function in the form of emergent processes of epigenesis of the organism.

Here are general and basic biological phenomena and processes which have a straightforward holistic collation on the basis of nonconventionally statistical application of quantum theory. The above list can be extended considerably to bear upon a host of other, biological phenomena which have previously appeared to be disconnected.

In Chapter 7, a prediction is discussed concerning the existence of specific, additive, long-range, intermolecular forces which emerge under the peculiar ("non-Gaussian") phenomenological conditions of molecular motion in many living processes.

References

Child, C. M. 1907. *Biol. Bull.* **12**: 89.

———. 1907. *Biol. Bull.* **12**: 177.

Halperin, W., and Wethirwell, D. F. 1965. *Science* **147**: 756.

Maximow, A. A., and Bloom, W. 1930. *A Textbook of Histology*. Philadelphia: Saunders.

CHAPTER 7

LONG-RANGE FORCES

Long-range, specific, additive forces emerging under non-Gaussian conditions of molecular motion in biology

Successful application of physical theory in biology should lead to useful predictions.

A prediction may be made that non-Gaussian conditions of molecular motion have an expression of critical significance to biology in the generation of long-range, specific, additive forces. These forces have the required characteristics for bridging the gap from protein and nucleic acid molecules to macromolecular aggregates or bions, to combinations of bions in cells, and to combinations of cell forms or tissues in organisms.

Hoffman (1957) has stated, "One speaks of long-range and of short-range forces depending on the phenomenon concerned, but the basic problem is how forces come into being that move parts of a living gel relative to one another. The movement is not merely a random wiggling; they are precise and the forces producing them have special properties. The observed biological phenomena at times point to forces not yet measured or studied in molecular physics."

Long-range forces very likely play a grand role in living form and function and undergird, in necessary ways, many of the outstanding phenomena of genetics, embryology, physiology, pathology, immunology, and pharmocology. Examples of biological phenomena and processes which may be dependent in at least some important manner upon long-range, specific forces are enzyme function, blood coagulation, antigen-antibody systems, host-invader (viral, bacterial) specificity, tissue identification, cell locomotion, cell orientation with other cells, vicarious calcification, collagen-vascular diseases, congenital malformations, and many degenerative disorders of macromolecular complexes, including atherosclerosis, arterioscleriosis, arthritides, and demyelinating processes. Before discussing these forces in detail, we shall give a short historical account of similar forces and their treatment.

FORCES ARISING FROM INDUCED DIPOLE-INDUCED DIPOLE INTERACTIONS OF FLUCTUATING ELECTRONIC CHARGE IN MOLECULAR OSCILLATORS WITH FINITE ZERO-POINT ENERGY

London (1937, 1942) treats forces arising from induced dipole-induced dipole interaction of fluctuating electronic charge. These are one of the forms of van der Waals forces. In London's treatment the forces arise from quantum mechanical coupling at ordinary temperatures of the oscillators which have finite zero point energies.

These forces fall off rapidly with distance (depend on the 7th power ($f \sim 1/r^7$) for simple molecules). His treatment requires knowledge of the microscopic electronic polarizability, a quantity not directly measurable. Also, it must be assumed for the treatment of condensed media that the force be additive (or nearly so), i.e. the attraction of a molecule A with molecule B does not appreciably affect its attraction to another molecule C.

Derjaguin, Abrikossova, and Lifshitz (1956) and earlier Casmir and Polder (1948) have treated the same force from the standpoint of quantum electrodynamics.

In this treatment the molecules are pictured as dipole "antennae" radiating and absorbing virtual photons. This interaction gives rise to a net force of attraction whose calculation needs no unmeasurable quantities. An essential feature of this treatment is seen to be the requirement that the distance of attraction be on the order of one quarter wave length of the virtual photon. This is to say that the radiation must arrive in correct phase (90°) for maximum interaction. Also the force is not assumed to be indefinitely additive ("and is in fact found not to be" [Derjaguin, Abrikossova, and Lifshitz]). The force then varies as the 7th power at short distances and as the 9th power at $r \geq \lambda/2$ for simple vibrational modes of molecules or atoms (where r is the distance and λ is wave-length).

M. Vold (1954) has extended a combined London-Casmir-Polder treatment to molecules having finite dimensions and many oscillators. Here again the electromagnetic retardation becomes important, i.e. the total attraction forces exerted by a set of oscillators in depth does not increase indefinitely since some of the radiation arriving at the interacting oscillator must be out of proper phase.

On the other hand, macromolecules having a large number of oscillators give a total attractive force due to electronic polarization at moderate distances (total energy of attraction $10kT$) which varies thus:

spheres as r^{-1}
rods as r^{-2}
platelets as r^{-3}

Jehle, Jos, and Bade (1959) have extended or generalized the electronic polarizations of London ($E \gg kT$) to include those oscillators in the "classic" region ($E \ll kT$) (where E is energy, k is Boltzmann's constant, and T is absolute temperature). One grave omission seems to be oscillators with $E = kT$, i.e. the vibration-rotation modes in the infrared region. Proteins have a large number of those modes with very specific orientations. The van der Waals forces can be considered to be very specific in that the molecules possess appreciable polarizabilities over a wide range of vibratory modes.

It would be of obvious advantage to extend Lifshitz's treatment to long wavelength radiators. This would give cognizance to forces with components which fall off slowly up to very long distances by analogy with Vold. The immense number of different kinds of oscillators with corresponding orientations (giving rise to polarized electromagnetic radiation) would contribute to the specificity as shown by Jehle. A corresponding advantage might be the ability to handle the non-indefinite additivity occurring in molecular aggregates.

BEHAVIOR OF LONG-RANGE, SPECIFIC, ADDITIVE FORCES IN A BASIC BIOLOGICAL UNIT, THE BION

Using the treatment afforded by Jehle, Vold, and Lifschitz, without rigorous mathematical extension, we are able to add several considerations to behavior of these forces in a new basic biological unit, the bion. The smallness of the volume, the largeness of the molecular sizes, the smallness of the molecular numbers, and the limitation of time intervals are factors which combine to create the special stiuation of biology. Each of the above treatments of van der Waals forces has its roots in the uncertainty relation. In this regard we may examine explicitly energy and time. From the uncertainty in measurement of these conjugate variables there arises a definite transition probability characterizing the exchange energy between two oscillators. (We say that within a certain time, on the order of the vibration period, we cannot tell with certainty which molecule has the energy.) Implicit in the uncertainty relation, though, is the assumption that the energies of many oscillators be distributed over a normal or Gaussian error curve. In this case the equality holds in the uncertainty relation: $\Delta E \cdot \Delta t \geq h/4\pi$. This gives rise to corresponding values for the transition probabilities and hence certain energies of interaction. Now we consider the biological situation where there are only a small number of each kind of oscillator. Clearly, the distribution of energies is not Gaussian and hence the uncertainty of action (energy and time) is greater than h by a certain factor, say $\in$, depending on the number of molecules. As the number of oscillators decreases the force per oscillator increases because of a greater transition probability. Figure 9 illustrates this concept schematically. In the figure, line A represents an individual van der Waals force component for a large macromolecular array.

The existence of directionalizing, long-range bonds in biological processes does not eliminate the need for and important operation of short range bonds of the covalent, ionic, and hydrogen types or even of shorter length van der Waals bonds.

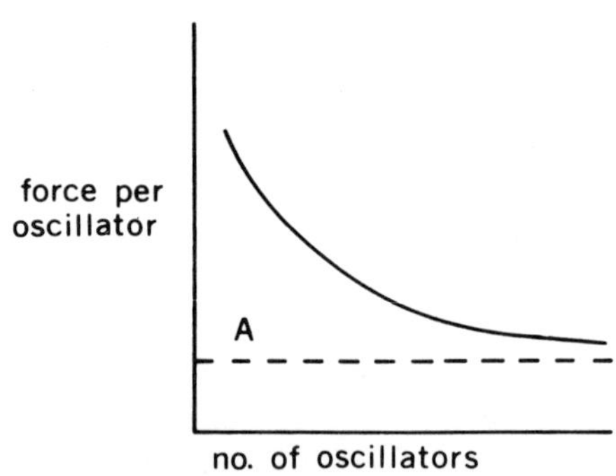

Figure 9. Behavior of the dispersion forces for low numbers of oscillators.

LIMITS OF THE SIZE OF NON-GAUSSIAN MACROMOLECULAR COMPLEXES

Under continued impact by random thermal motions of the solvent or suspended molecules there may occur an interval during which the one-sided impacts continue long enough to drive a given macromolecule away from the complex. Here we are describing a situation which depends on the accumulation of random high energy collisions on one side of the macromolecule (or part of the complex) within a short time period. This thermal disruption within a certain time, combined with the fall-off in total force build-up with the aggregation number of the macromolecules, would give a range of maximum size of a complex that could be formed.

One has here the same sort of probability consideration as that when calculating the total number of tosses of a coin to "assure" a certain number of heads in a row. One considers the overall time interval in which enough directed impacts of sufficient energy occurs within a given short time. If the aggregating forces are large, for example $20kT$ per molecule, one would expect an extremely long time necessary to assure ten high energy impacts (each of $2kT$) in a row to one side of the complex. On the other hand, if the forces are weaker, for example $10kT$, then a much shorter time would be necessary for 5 directed impacts to occur with 1/5 the necessary energy ($2kT$). Burnside (1959) has shown the probability (or time interval) to be a power function strongly dependent upon the number of events required to happen in a row.

INTERPRETATION OF FUNCTION OF PROTEIN PORTIONS OF ENZYMES AND ENZYME CATENA

Having presented a discussion for the plausibility of these forces, we may nicely describe one further aspect of biological function which to date has remained rather amorphous. The function of the protein portion of an enzyme has been relegated to the role of providing a skeleton for the "active site." We would propose that the rest of the protein provides a very large surface area for impact of solvent molecules possessing sufficient kinetic energy to initiate reaction. The long-range coupling forces existing between the vibratory modes now can be seen to provide a "direct-access" communication route favorable for transmission of activation energy between the protein surface and active center. Such an increase in effective surface area of the active site would greatly increase the rate of reaction compared to the situation in which the small site of attachment of substrate would be the only receiver of an impact sufficient to cause reaction. A further extension could now be seen in similar action of an enzyme in a catena. Energy of reaction instead of being degraded to heat could be made available over intermolecular coupling to serve as activation for a later reaction. At this point, it is appropriate to intimate that such an enzyme acting in virtual isolation or in great number in a beaker may display an immense quantitative and qualitative deviation from its behavior in an aggregate.

EXPERIMENTAL HINTS OF AND PHENOMENA MOST EASILY EXPLAINED BY SPECIFIC, LONG-RANGE FORCES IN BIOLOGY

The series of careful experiments of Rothen (1948, 1950, 1959, 1962) give direct evidence of the existence of long-range forces. He has, to be sure, evidence for the periodicity of long-range forces which act between protein layers which are separated by films.

Enzymatic Reactions at a Liquid-Solid Interface

Rothen set up the experimental conditions which are illustrated in Figure 10 to test enzymatic reactions at a liquid-solid interface. A protective blanket of barium stearate or formvar did not prevent the enzymatic reaction from taking place (trypsin on albumin); remaining albu-

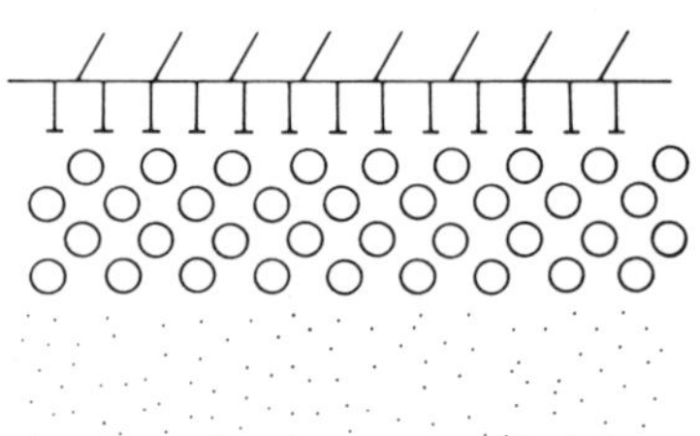

Figure 10. Experimental demonstration of long-range action of trypsin on protein (albumin) through many layers of amorphous material (after Rothen, 1959).

min was detected by antibody. The minimum thickness of a plastic blanket required for protection was a function of the number of layers of adsorbed protein. A formvar layer 20 Å thick was sufficient to protect one layer of bovine serum albumin, whereas it took 600 Å of formvar to protect six mono layers. The tryptic action also depends on the number of fatty acid layers below the protein layer (toward the slide). Trypsin can be forced through a layer of formvar 1000 Å thick if the field is strong enough, but cannot go through even 50 Å if weak. All tested proteins can force trypsin through the blanket. RNA has no effect on forcing trypsin through the blanket. The force seems to be periodic, especially with the thickness of the underlying substrate (anchorage).

Antigen-Antibody Precipitation Through Several Layers of Fatty Acid

Rothen tested for residual albumin in the above experiments by applying prepared antibody which was precipitated through a number of layers of fatty acid or formvar. See Figure 11.

Long-Range Interactions Between Cell and Substrate

Rosenberg (1962) has discussed the relative roles in long-range interactions of "long-range forces" and powerful "short-range forces" that operate by successive polarization of neighboring molecules over considerable distances. He remarked that serum fractions, surface charges, basic peptides, divalent ions, electric double layers, van der Waals forces, etc. have been considered, but mainly from the view point that the interaction properties of a substratum reside only in its surfaces. Physical thickness and bulk properties of the substratum have not ordinarily been regarded as significant. The only variables in Rosenberg's

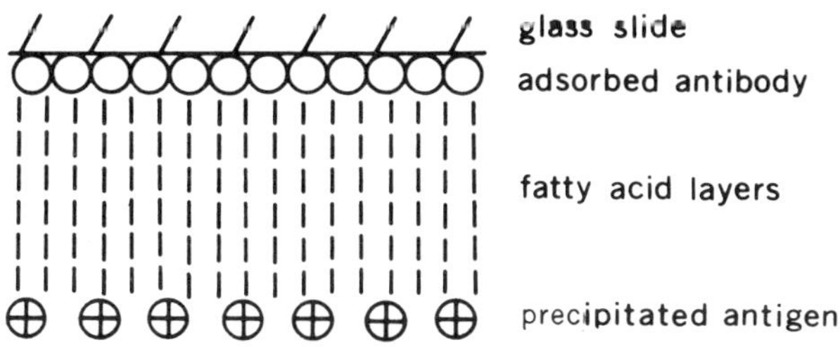

Figure 11. Long-range antigen-antibody precipitation through multiple layers of fatty acid, after Rothen, 1959

experiments were the number of monolayers (laid out in "staggered" fashion) and the type of substratum or supporting base, yet the rates of attachment and spreading of human conjunctival cells exhibited indisputable dependence on both variables. The problem of incompleteness of monolayers and the "ripping reaction" of Sher and Sobotka (1955) could not account for the anomalous results, so less orthodox explanations relating to long-range reactions must be considered.

Specific Pairing of Chromosomes

Specific pairing of chromosomes would be most easily explained by long-range, specific forces (Hoffman, 1957). Although submicroscopic attractiveness by intermolecular forces may be inferred from substrate-enzyme and antigen-antibody interactions, the specific pairings of chromosomes in early meiosis and of giant diploid chromosomes of flies (Diptera) are directly observable under the microscope, as is also re-aggregation of cells in sponges and embryos. The periodicity or reversal of such forces is attested to by the repulsion between duplicate or homologous chromosomes in mitosis. Electrostatic and other short-range forces are not implied to be nonoperative; the additive, long-range, specific forces offer an over-arching or directionalizing influence. Ions in a gel are not sufficiently long-range (fractions of a micron) to be critical here; the gegion (opposing positive and negative ion) effects must also be considered with their screening influence on chromosomal change.

Reaggregation of Cells

Though the numerous experiments described by Weiss (1959) and Mascona (1961) do not exclude other explanations, they certainly are suggestive of the interaction of proteins pictured here.

Weiss and Mascona described the reaggregation of like cells of many different kinds of embryos, sponges, everted hydras, mammalian organs, etc. which had previously been separated into heterogeneous cellular suspension by various means. Long-range, specific, additive forces could account elegantly for these phenomena. The reaggregations of cells of like types occurred with remarkable (and seemingly inexplicable) regularities in time. This time dependency would be quite consistent with our thesis of required limitation in time and space of the number of accessible degrees of freedom in order to have indeterminacy of molecular motion.

Tissue Freezing Across Insect Carapace

Supercooled (20° C. or so) insect tissues (whole, under the carapace) do not freeze until water is dropped onto the carapace (Salt, 1963). (See Figure 12.)

The carapace of the cockroach is very impermeable to water, so nucleation of ice outside the carapace has (*a*) changed the local surface of the carapace, and this is reflected below, or (*b*) residual long-range forces from protoplasm below the carapace are influenced. Nucleation is usually a molecule-to-molecule interaction starting from the lower energy surface of an incipiently forming ice crystal. The macroscopic thickness of the carapace would seem to provide a barrier to this propagated short-range interaction. The possibility of long-range forces must be considered.

Similar Orientation of Sperm Tails in Spermatogenic Cysts

The sperm tails of many submammalian species are indistinguishable from simple flagella. The micrograph shown in Fig. 13 presents transverse sections of a bundle of sperm tails in the lumen of one of the spermatogenic cysts of a fish testes (*Lebistes reticulatus*). Although the cross sections each belong to a separate spermatozoon, the central pairs in all the tails are oriented similarly just as are those of the cilia of an epithelial cell. No clear explication has been forthcoming for this orientation. Long-range forces would account elegantly for this orientation.

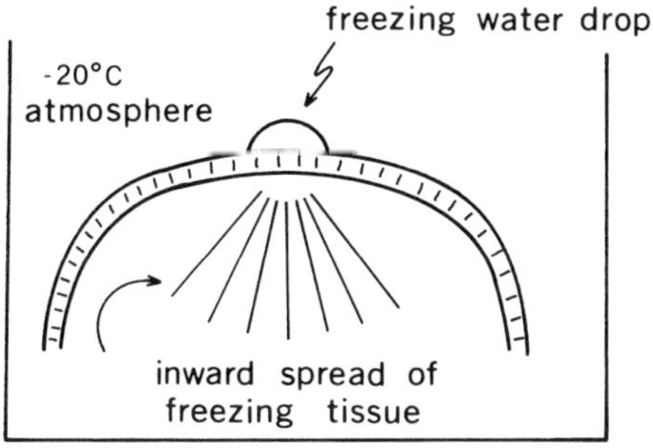

Figure 12. Freezing of supercooled insect tissue through thick carapace by placement of drop of water. This is an example of possible long-range forces between incipient nucleation crystals of ice and protoplasm below carapace (after R. W. Salt, 1963, Canad. J. Entom. **95:** *1190).*

Synchronization of Wave Movements of Microscopic Undulating Organisms

Synchronization of wave movements of two undulating organisms such as spirochetes or spermatozoa occurs when they come into close proximity (Cleveland and Grimstone, 1964; Gray, 1928 and 1951; Holwell, 1966). Also, in dilute suspensions of bull sperm where the av-

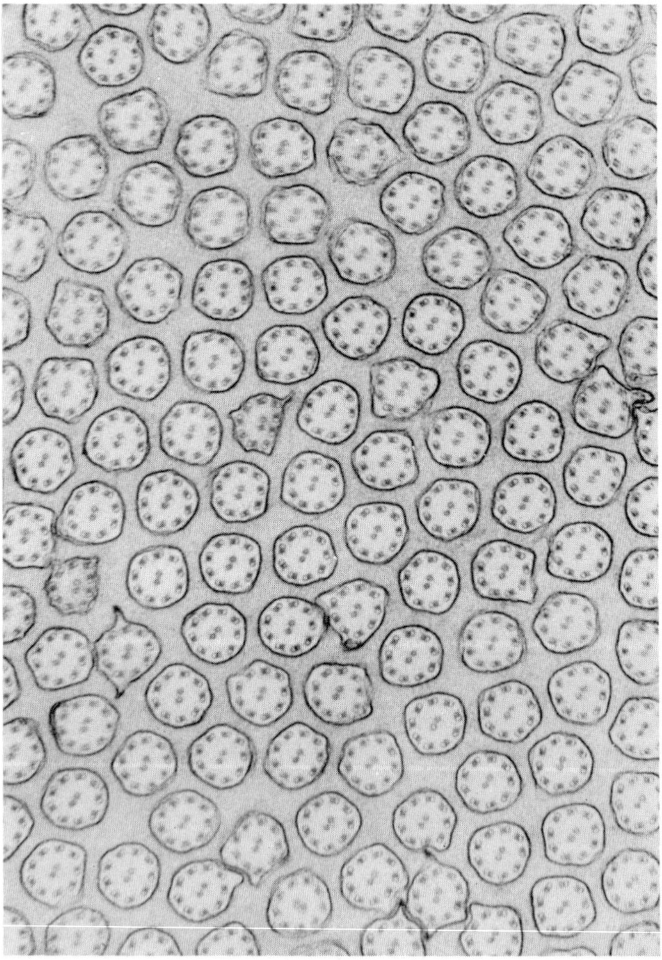

Figure 13. Electron micrograph of transverse sections of a bundle of sperm tails in testis of fish (Lebistes reticulatus). *Each cross section is a separate spermatozoon. The central pairs of fibrils in all of the tails are similarly oriented. The operation of long-range forces is a strong possibility. (Collidine-buffered glutaraldehyde and osmium fixation. Lead citrate staining.* (From D. W. Fawcett. 1966. *An Atlas of Fine Structure*. Philadelphia: W. B. Saunders Co.)

erage distance between adjacent flagellated sperm cells is relatively large, a weak repulsive interaction has been reported (Rikmenspool, 1962) that prevents synchronization. This periodicity of attraction and repulsion at varying, close distances is what would be expected of long-range forces. Viscomechanical effects have been postulated by Jahn (1964) for this phenomena, but it has a more plausible explanation in terms of long-range forces of the nature here described.

Synchronized, Communal Motility in Slime Bacteria, True Bacteria, and Slime Organisms

A swarm stage occurs in slime bacteria (Myxobacteriales) with communal motility without external cilia or flagella. This is reminiscent of motion in *Cystophaga* of the Spirochaetales, in sulfur bacteria (Beggiatoaciae), in diatoms (*Navicula*), in algae (*Oscillatoria*), in Cyanophyceae, in iron bacteria, and in true bacteria. The true bacteria, *Bacillus alvei* and *B. rotans*, express motility in a slowly milling mass of bacteria which synchronize their motion and may produce dextro- or levorotory discs (Lamanna and Mallette, 1965; Pollard, 1963). The Mycetozoa, or slime organisms, at the swarm stage (Wood, 1958, 1964) are a mass of living protoplasm in which are seen many nuclei. They have an amoeboid-like movement for several days in the single, living plasmodium or swarm stage and then cease motion as reproduction beings. Stalks are sent up with sporangia (sporebearing cysts) on their tips. Each spore is a nucleus with a cellulose wall. The wind disperses the spores into water where germination of each spore leads to formation of a naked, ameba-like creature, which develops a flagellum later. These forms rapidly multiply by fission, and after this the flagella are lost. Thereupon, a general coalescence takes place to form the multinucleate swarm stage again with communal motility.

These organismal motions are too complicated for complete description by classical mechanism. Long-range, directionalizing forces among bions and macromolecular complexes seem to be an unavoidable part of the explanation in view of the indeterminate nature of the constituent molecular motion.

Paradox of Maximum Viscosity of Bacterial Suspensions at Isoelectric Point

Bacteria in a suspension exhibit maximal viscosity at the isoelectric point. This unexplained phenomenon is most curious in that colloidal solutions have minimal viscosity at or very near the isoelectric point. Bacterial sizes range in the upper limits of sizes of colloids and collectively in suspension demonstrate many properties of colloidal solutions, such as light scattering (Tyndall effect), Brownian movement, viscosity increase of suspending medium, magnetism in electric fields, agglutination by salt or dehydration effects.

The absence of Brownian movement within living cells and maximal viscosity of bacterial suspensions at the isoelectric point both point to a fundamental fact, namely, the basis of the physical nature of protoplasm is not encompassed by the laws of ordinary chemistry. Long-range forces come immediately to mind, but offer little advancement in scientific explication without some appropriate and nondualistic theoretical grounding. Our thesis of nonconventional statistical application of quantum theory explicitly provides this grounding.

Maximal Reaction Rates in Large Molecule Synthesis Indicating Specific Orienting Forces

Theoretical calculations of the maximum reaction rates involved in large molecule synthesis of various kinds have led Pollard (1963) to the conclusion that a force with somewhat the same type of characteristics that we have described must be operative in these processes. The conclusion is that to achieve the required order of magnitude of synthetic rate one must have specific and orienting influences operative in the assembly of large molecules.

Phenomena of Host Specificity

Much has been written about host-viral, host-bacterial, host-foreign body, and host-transplant specificity. The modern surgical effort in organ transplantation began many years ago with such surgeons as Carrel and has made slow progress (except for transplantations in identical twins). The surgical technics of transplanation are not the basic impediment to progress; rejection phenomena are the major impediment. What is needed is new insight into the nature of biological processes. This brings the discussion to an important point.

The infectivity for certain cells and shedding of protein components of viruses and bacteriophages with entry of nucleic acids into cells provide hints that long-range forces are operational with additivity, specificity, and both attractive and repulsive characteristics, i.e. definite periodicity or distance dependence. The additivity, specificity, and long-range of such forces is quite plausible in view of the relatively large number and specific orientation of vibrational-rotational modes of oscillation in the infrared (with $E = kT$) in association with protein molecules.

The "stickiness" of tumor cells (Wood, 1958, 1964) may be correlated with this thesis, especially in lymphoid or reticuloendothelial tissues (blood and lymphatic vessel walls are not to be forgotten) which are composed of primitive cells with more variety of information, offering more possibilities for alignment. The separation by the spleen of degraded red blood cells from the passing blood is another example, as is the loss of stickiness of cells within transplanted tissues and conversely the adhesiveness of host phagocytic cells for cells of transplanted tissues. This is one form of expression which transplant rejection takes under the special biological conditions of quantization.

Mitogenic Radiation

Indirectly, mitogenic radiation may be involved with the operation of long-range forces in living protoplasm. This subject is taken up in greater detail in the next chapter.

References

Burnside, W. 1959. *Theory of Probability*. New York: Dover.

Casmir, H. B. G., and Polder, D. 1948. *Phys. Rev.* **73**: 360.

Cleveland, L. R., and Grimstone, A. V. 1964. *Proc. Roy. Soc. London Sec. B* **159**: 668.

Derjaguin, B. V. 1960. *Scient. Amer.*: **203**: 47.

Derjaguin, B. V., Abrikossova, I. I., and Lifshitz, E. M. 1956. *Quart. Rev.* **10**: 295.

Gray, J. 1928. *Ciliary Movement.* London: Cambridge University Press.

———. 1951. *Nature* **168**: 929.

Hoffman, J. S. 1957. *The Life and Death of Cells.* Garden City, N.Y.: Doubleday.

Holwell, M. E. J. 1966. *Phys. Rev.* **46**: 696–785.

Jahn, T. L. 1964. *Am. Zool.* **4**: 264.

Jehle, H., Jos, J. H., and Bade, W. L. 1959. *Proc. 1st Nat'l. Biophys. Conf.* New Haven: Yale University Press.

Lamanna, C., and Mallette, M. F. 1965. *Basic Bacteriology*. Baltimore: Williams and Wilkins.

London, F. 1937. *Trans. Faraday Soc.* **33**: 8.

———. 1942. *J. Chem. Phys.* **46**: 305.

Mascona, A. A. 1961. *Scient. Amer.* **205**: 143.
Pollard, E. C. 1963. *J. Theoret. Biol.* **4**: 98.
Rikmenspool, R. 1962. In *Spermatozoan Motility*, ed. W. Bishop, p. 31. Washington, D.C.: American Association for the Advancement of Science.
Rosenberg, M. D. 1962. *Proc. Nat'l. Acad. Sci.* **48**: 1342.
Rothen, A. 1959. *Proc. 1st Nat'l. Biophys. Conf.* New Haven: Yale University Press.
———. 1948. *J. Am. Chem. Soc.* **70**: 2732.
———. 1950. *Science* **112**: 530.
———. 1962. *J. Coll. Sci.* **17**: 124.
Salt, R. W., 1963. *Canadian J. Entomol.* **95**: 1190.
Sher, I. H., and Sobotka, H. 1955. *J. Coll. Sci.* **10**: 125.
Vold, M. 1954. *J. Coll. Sci.* **9**: 451.
Weiss, P. 1959. In *Biophysical Science*, ed. J. L. Oncley, New York: Wiley.
Wood, S., Jr. 1958. *Arch. Path.* **40**: 92.
———. 1964. *Bull. Sw. Acad. Med. Sci.* **20**: 92.

CHAPTER 8

BIONIC RADIATION

Coherent, laser-like radiation in the infrared from bions, involving mainly vibrational-rotational energy transitions in molecules

The prediction of bionic radiation follows from a potential for exchange of virtual photons among limited numbers of molecules in small spaces and times, with the molecules undergoing open exchange of mass and energy with the environment at ordinary temperatures and pressures. This potential represents symmetry properties between force fields (or long-range "exchange forces" of bions) and large molecules of the bions which generate the fields. The symmetry properties are those of energy input and output that manifest themselves with a change of (energy) state of bions. Confirmation of the existence of bionic radiations would lend credence to the thesis of this monograph.

Because of the virtually point nature of the radiation sources and the large number of vibrational-rotational energy transitions in molecules that can be made to oscillate in bions, bions may be anticipated to produce coherent, laser-like radiation mainly in the infrared. Some of the wavelengths generated will overlap into the visible and ultraviolet ranges, in view of the density of unsaturated bonds in the molecules, and into the far infrared and microwave ranges, in consideration of the strong theoretical possibilities that have been discussed for quantization of translational motion of molecules in bions. This "bionic" radiation may be of an anomolous electromagnetic type or of an entirely different (new) variety.

REPORTS OF RADIATIONS FROM LIVING CELLS, ORGANISMS, AND COMPONENTS OF LIVING CELLS AND ORGANISMS

The Gurwitschs' (1923a, 1923b, 1925, 1932; see also commentary by Rahn, 1936, pp. 54, 55) investigated rather intensively a phenomenon in which the growing tip of onion root, placed at a noncontacting distance and at an angle toward the side of a second growing tip of onion root, resulted in an increase in number of cells undergoing mitotic division in the tip of the second (detector) onion root in the region nearest the tip of the first (emitter) onion root. (See Figure 14.) If the tip of the emitter onion root was heated to 60° C. prior to exposure, there was no increase in mitosis in the tip of the detector root. The term mitogenetic or mitogenic radiation was adopted for this unusual action at a distance. This radiation came also to have the name of Nancy or *N* rays, after the town in France where it was first observed from contracting muscle by Blondlot and Charpentier (see Harvey, 1940, pp. 221–23).

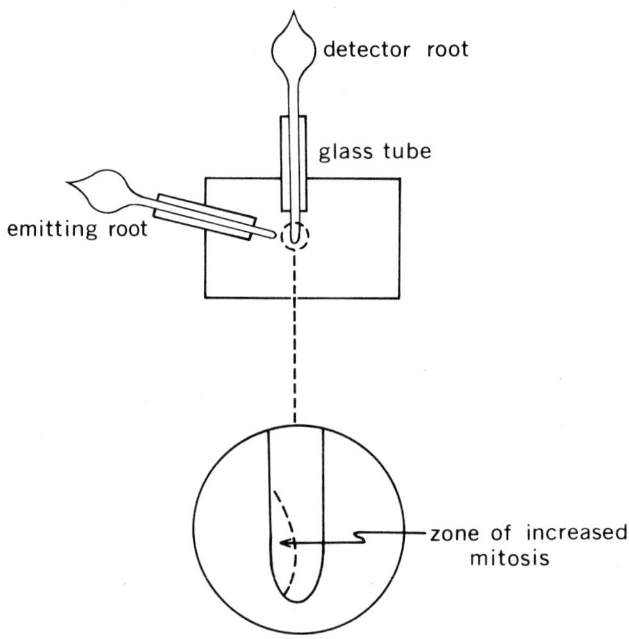

Figure 14. Gurwitschs' experiment with growing onion root tip as emitter of mitogenetic radiation. The root tip is at an angle to a second growing root tip which is the detector of the hypothesized radiation, the presence of which is manifested as an increase in number of detector tip cells that are undergoing mitotic division.

This statement concerning "mitogenic radiation" is introduced at the outset in order that a circumscribing statement be may made about "bionic radiation." The two are not intended automatically to be synonymous. The prediction of bionic radiation comes from the plausibility hypothesis of this monograph concerning quantal phenomena in living cells and organisms. Mitogenic radiation has at least some of the properties that would be predicted for bionic radiation and may be bionic radiation, but no claim is made in this writing that one is the other.

With this understanding, a recounting may be made of reported invisible radiations, similar to mitogenic radiation, between living beings of various sorts.

The monograph of Rahn in 1936 is a rich account of previous work. He found the growth rate of yeast and bacterial cells to be very sensitive to mitogenic radiation. He noted that the radiation appeared only from living organisms in quartz vessels, and concluded that it was ultraviolet radiation. He observed that a similar emission from oxidative reactions and proteolysis was enhanced by diffuse daylight. He placed the wavelengths at 1800–2600 Å. His explanation for the emission was "leak" of activation energies. For the growth rate of bacteria and yeast to be stimulated, the intensities of radiation had to be moderate and the wavelengths had to be below 2600 Å. High intensities failed to enhance or retard growth. The distance between emitter and receiver was about 2–3 cm. in air, increasing to 10–15 cm. with intermittent radiation.

The following tissues were found to be good radiators:

cells in culture
corneal epithelium of frog
sea urchin eggs
brain, blood, active muscle in most adult animals

Intermittent, rhythmic radiation seems to have had more effect than continuous radiation. The rate of conduction of mitogenic radiation in frog sciatic nerve by secondary radiation was 30 meters/second.

Muscle radiated during work and when greatly fatigued. Blood, outside its blood vessel, lost its radiating power very quickly. Myelin, but not skin (cornium), was found to be transparent to radiation. Lecithin films scattered the radiation. The lag phase of bacteria and yeasts was the phase of maximum response. If the base of onion root was severed or narcotized, the tip did not radiate. Irradiated sea urchin eggs developed into abnormal larvae; this seems to have involved stimulation of mesenchymal growth with distortion of larvae.

Glass absorbed the radiation. Quartz did not. Glass with a layer of paraffin did not. Electrical insulators did not prevent radiation passage. Electrical conductors did prevent passage. Other physicochemical detectors of mitogenic radiation are reported to be: Liesegang rings, decomposition of hydrogen peroxide, flocculation of colloids (gold sols), and photoelectric counters (Metcalf and Quickenden, 1967; Troitskii et al., 1961). (Modern photoelectric counters have senstivities of 10^{-11} ergs/cm.^{2}/sec.) (See Figure 15 for effect of radiation on Liesegang rings.)

The following observations have been reported concerning this radiation from living sources (Rahn, 1936). (*a*) Very few diseases except cancer and (perhaps chronic tonsillitis) prevent blood radiation. (*b*) Wounded leaves of plants and wounded tadpoles radiate; hence, the implication is that mitogenic radiation occurs in wound healing. Standard wounds in tadpole tails healed more rapidly when exposed to radiation from tadpole pulp. Maximum stimulation was observed at two times after wounding: immediately to 24 hours, and at 48 hours. (*c*) Bacterial growth was retarded by radiation from another culture. (*d*) Growth rate of yeast was retarded when the culture was irradiated with menstrual blood. Yeast showed enormous vacuolization, growth of hyphas, and giant cell formation. Some bacteria were killed, others showed abnormal forms, some were not affected by irradiation from menstrual blood. (*e*) Malignant tumors radiated strongly, and benign ones only weakly. (*f*) Blood from patients with the malignant tumors decreased mitogenic radiation from blood. (*g*) Necrotic parts of malignant tissue showed proteolytic radiation spectrum (see below). The main growing body of malignant tissue emitted glycolytic radiation. The conclusion from *f* and *g* by Rahn was that "growth stimulating source of radiation is removed from the blood and concentrated in the tumor." Another conclusion was that cancer is frequent in old age when blood radiation is lower.

Sources of mitogenic radiation can be: contracting muscle, oxidations of biological compounds or pulps, sugar fermentations, proteolysis, nuclease (phosphatase) activity, phosphate cleavage, etc.

Rahn gave these values for radiant wavelengths of mitogenic radiation:

Oxidations: (pyrogallic acid by air) (glucose by Mn O_4) (serum by H_2O_2)	2200–2280, 2280–2340 Å
Sugar Fermentation:	1910–1920, 1930–1940, 2120–2180, 1950–1960 Å
Nuclease (phosphatase):	2150–2160, 2240–2250, 2280–2290, 2350–2360, 2460–2500 Å
Phosphate Cleavage:	2010–2020, 2030–2060, 2090–2110, 2120–2130, 2140–2150 Å
Proteolysis	1980–1990, 2030–2050, 2110–2130, 2300–2310, 2340–2350, 2380–2400, 2410–2420 Å

Rahn observed that the detectors had to be correct as to biological phase. For example, the most sensitive detectors in his work were bacteria and yeasts in the lag phase of growth of the cultures.

In 1940, in a book called *Living Light*, Harvey gave a summary of detectable radiations from living materials, and criticized the observations on mitogenetic radiation on several major points. He stressed these radiations were not of the ordinary electromagnetic variety. He warned against studying such anomalous phenomena which challenge instrumental and technicological capabilities. He noted that the anomalous radiation came from many different things, including ordinary chemical oxidation reactions. He recounted that in certain experiments the same results were obtained with or without a prism in the spectrographic setup. He concluded that mitogenic or mitogenetic rays were probably the result of uncritical investigation of insufficiently controlled experimental materials with which it is difficult, if not impossible, to work.

A discussion of Harvey's criticisms may be helpful. In 1940, scientists were not acquainted with coherent radiation from extended sources. Coherent radiation is different in many of its manifestations from ordinary or incoherent electromagnetic radiation. Distance dependence of effects very near the source, and directionality of effects can be completely different for the two types of radiation. For example, coherent radiation may be diffracted by a slit alone, possibly explaining some of the spectroscopic experiments. Furthermore, this radiation could, by virtue of its source, have a very specific spectrum. The degree of intermeshing or fit of source and detector could lead to a range of partial or incomplete effects. Mitogenetic radiation seems to be emitted and detected primarily by living systems. This leads to unusual difficulties in manipulating the experimental arrangements. (*See* p. 129.)

Regardless of spurious examples of radiation from living material and of the many different substances which supposedly produce such radiation, the really significant point is that there is even a single instance of anomalous radiation from living material which has been or can be detected.

The problematic nature of the technology for detection of anomalous radiation from living matter makes such radiation more significant in view of this monograph's plausibility argument for its mechanistically indeterminate basis. The anomalous behavior of this radiation is all the more reason to study it.

Although many experiments were carried out in the attempt to detect radiation from living material or components of living material, the *nature* of the radiation was not critically studied. In particular, its dependence on distance was not investigated systematically. The wavelengths were also in question. The aspect of superposition of waves in or out of phase (coherence) was not considered with respect to the source of radiation.

The radiation could not all have been of ultraviolet type, since it penetrated paraffin on glass. (The projection from this manuscript would place the radiation most plausibly in the infrared of lower frequency).

Harvey did not remark about or appreciate the implication of the wave phase or coherence problem in radiation. Radiators are incoherent in ultraviolet at long distances in comparison to size of the radiator. Alignment of wave phase in radiation would be very important in radiation from oscillators in bions. We suggest that coherence may have been overlooked by the investigators of mitogenic radiation and by Harvey. All of the typical examples of ultraviolet radiation in the past have been incoherent and are today with the possible exception of lasers.

COMMENTS AND PROJECTIONS FOR FUTURE INVESTIGATIONS OF BIONIC RADIATION

The vital importance of coherence or incoherence of bionic radiation would be expected to show up in certain aspects of the radiation between bions. For instance, the condition of coherence of radiation would have a distinct influence upon the *distance dependence* of the radiation between emitter and detector. This is demonstrable in that reduction of distance might be associated with a lesser force, or no force, for a certain distance. This could be secondary to phase difference of radiation wave rather than intensity difference, especially if the radiation were of fairly long wavelength. Certainly, attempts to repeat detection of such previously detected radiation would be fraught with difficulties, and in the past could have created some of the disparities in the different investigations of mitogenic radiation. The investigators of mitogenic radiation did not ask the question, "Is the anomalous character of this radiation due to coherence which is not being taken into account?" How atoms are lined up in molecular radiation is important. All the spin states must be pointing in the right direction to get coherent radiation. It will be important in any future work to consider coherence.

There are now four well recognized types of forces in physics. From these four types of forces that might be said to exist in actuality, there are four types which might be said to exist virtually. The latter represent some manner of symmetry or reciprocity property between force fields and particles or processes which generate the fields (Weisskopf, 1965).

In the case of an atom, an electron is kept around the nucleus by an electric force or field, which may be pictured as originating in the *virtual* exchange of photons between nucleus and electrons. However, exchange of an *actual* photon from an excited or emitting atom to a receiving atom occurs with several simultaneous changes. As the photon is emitted, an excited electron falls to a lower energy level in the stable state of the emitter atom. As the photon is received, an electron goes to a higher energy level in the stable energy state of the receiver or detector atom.

Likewise for the nucleus, a meson force field is thought to hold the nucleus together as a virtual exchange between nucleons or particles of the nucleus. This potentiality for virtual meson exchange between nucleons gives rise to a field or force which keeps the nucleons in the nucleus. A real meson may be gotten by exciting a nucleon in isolation and forcing a meson (a real particle) to be emitted.

Returning to the electron, if an electron is excited, a real photon or electromagnetic radiation may be produced. One is familiar with this, and with the virtual exchange of photons between nucleus and electron, as the electric field. One has become familiar with other real radiation and fields (virtual radiation) in the nucleus under other appropriate names. We point to the symmetry (or reciprocity) between *fields* and *particles or processes which generate the fields* in the above cases as analogous to what we are attempting to describe for bions. The potential for exchange of virtual photons is involved in generation of long-range, specific, additive forces which help to keep microparticulates (large molecules) together as a bion. The existence of real radiative exchange between bions takes the form of those properties of energy input and output that manifest themselves with a change of energy levels of the stable states of the participating bions.

If one growing microbe is placed close to another, and each influences the other's growth, how can this be described? As an electromagnetic or mesonic or what kind of radiation? The problem in biology involves the same task which is encountered in describing electromagnetic or mesonic radiation. After the properties of the radiation are observed or investigated (which lies ahead for bionic radiation), the whole matter will take on much more of a unified aspect that may be susceptible to theoretical formulation.

To repeat for greater clarity, we interpret the long-range forces for bions as "exchange forces" which arise from uncertainty in position and momentum or time and energy for large molecular motion in essential living processes. These exchange forces help hold bions together. These interactions or exchange forces of large molecules in bions have a mechanistically indeterminate basis which involves exchange of virtual photons (bionic radiation). Note must be taken of the distinction that bions may also be excited, so as to emit photons (also bionic radiation). "Mitogenic radiation" may be bionic radiation of either type above.

The detection of exchange forces between ordinary matter is an extremely delicate operation, if a distinct measurement is to be made. Such measurements have been carried out by Russian and Dutch investigators (Bloch *et al.*, 1960; Derjaguin *et al.*, 1956).

The biological "unit" or bion may be acting as an amplifier of effects of exchange forces, an example being where one microbe's growth affects another microbe's growth.

The authors are interested in investigating the properties of bionic or mitogenic-like radiation in a systematic manner rather than merely attempting to detect further instances of the radiation. This will redound to the thesis of this monograph that many living processes are indeterminate in operation and quantal in description.

Despite experiments with spectroscopes, the nature of mitogenic radiation has not been investigated. The anomalies of transmission point up that the prism is not needed, and Harvey's criticism that the transmission occurs with or without a prism is an overlooked opportunity.

If Liesegang rings could be made into sensitive detectors, one side of the problem would be solved, i.e. the detector could be put in a vacuum or biologically hostile environment, which could not be done if the detector and emitter both must be biological. (Our preliminary experiments support interpretation of Liesegang ring formation as a dramatic macroscopic exhibition of quantized diffusion.) If a reliable non-biological detector can be discovered, bionic radiation could be investigated more definitely, that is operationally. Transmission through various media could be studied: glass, thin metal foil, mica (an ideal dielectric), vacuum, air, etc. The anomalous transmission goes through myelin. If this observation is correct, and since myelin is an extremely good scatterer for ultraviolet, this may help identify the wavelengths.

A non-biological detector would facilitiate investigation of dependence of bionic radiation on distance. It will be important to know whether the inverse square law of distance of radiation is obeyed. Any anomalous behavior such as fall-off or build-up with distance which is different from the inverse square law would indicate a phase or coherence problem. We suggest further experiments in which bionic radiation from onion root tip or bacteria is detected in the presence of Liesegang ring formation, as illustrated in Figure 15. (We have detected interference phenomena with Liesegang rings and growing onion root tips which have been separated so as to eliminate any material transmission between the two. The Liesegang rings were placed inside quartz cuvettes. The onion roots were placed outside the cuvettes at a distance of from two to four mm. from the rings. A slit in a metallic shield was used to collimate the radiation from the onion root tips.)

For the investigation of wavelengths of bionic radiation, interference

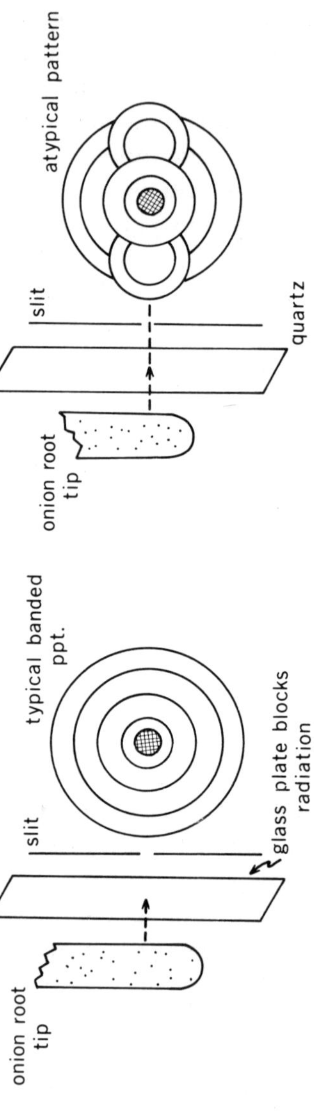

Figure 15. Effect of mitogenetic radiation of Liesegang rings (after Rahn, 1936).

wedges or interference filters may be used for determination of limits of wavelength in the infrared. If bionic radiation is not electromagnetic and is phase dependent, interference devices will not be helpful.

The point is to characterize bionic radiation itself, not just to list types of radiators or receivers. This may only be possible when further operational data are collected. If successful characterization of transmission properties of the radiation can be accomplished, it will be possible to study interposition effects on biological materials.

Experiments involving "solid-state carcinogenesis" may serve as prototypes for these interposition effects. A wide variety of solid materials including glass, noble metals, and plastics are able to induce tumors in rats upon subcutaneous implantation if the extent of the solid surface is large enough (Bischoff and Bryson, 1964). Conversely, implantation of equivalent amounts of the material in the form of perforated disks, fragments, or powders causes no tumors. It may be argued that the uninterrupted solid surface prevents the free interchange of tissue fluid. The accumulation or deprivation of substances whose presence or absence eventually leads to a malignant transformation is taken to be the underlying cause of the tumor. Two observations seem to vitiate this explanation, and the issue is by no means settled (see above reference). Selye (1961) found that when glass tubes were implanted, tumors now formed preferentially at the edges where interchange of fluid would be unimpeded. The second observation is that quartz disks are much less carcinogenic than glass under similar conditions (Bryson and Bischoff, 1968). Since quartz is much more transparent than glass to mitogenetic radiation and mitogenetic radiation exerts highly selective action on growth processes, an explanation of these interposition effects appears tenable. Protti (1934) found that suspensions of yeast were lethal to cancer cells in culture when the two were separated by quartz. Normal cells were not affected by like treatment. Should normal tissues depend, if only partially, on continual exchange of radiation to maintain a balance between growth, maturation, and senescence, interruption of this radiative interchange as described above could give rise to an incipient cancerous process in susceptible cells. Such an hypothesis could be easily tested once the transmission characteristic of materials were known in detail.

In summary, a plausible theoretical basis has been developed in this monograph for a type of radiation which we have termed bionic. Experiments have been conducted which have suggested some type of radiation from living organisms. This has been called mitogenetic radiation. Critics of the experimental work on mitogenetic radiation have viewed the radiation as a form of bioluminescence at the extremes of detection by sensitive instruments. Our theoretical considerations are compatible with a type of radiation which may include mitogenetic radiation. Bionic radiation may have an electromagnetic nature, but not necessarily so. If it is electromagnetic, it is of anomalous character. It is probably coherent in nature with peculiar distance-dependence. The wavelengths are probably mainly in the infrared. Some type of directed molecular motion is implied. If bionic radiation can be corroborated and is proved not to be of an anomalous electromagnetic nature, a new spectroscopy beyond the three kinds of Weisskopf (1968) may be in the offing. The implications of such a new spectroscopy for biology are quite promising because it might lead to methods for identification of tissues in organ transplanatation, of bacteria in infections, of cells in neoplasia, and of neural microstates involving synaptic connections associated with mental illness, as well as with aberrant cells or substances associated with aging, atherogenesis, collagen-vascular diseases, or malforming changes in embryogenesis.

In Chapter 9 further predictive aspects are discussed of the broader quantum interpretation of living processes with implications for pharmacology, physiology, and pathology of specialized organisms.

References

Bischoff, F., and Bryson, G. 1964. In *Progr. exp. Tumor Res.* **5**: ed. F. Homburger, p. 85. Basel/N.Y.: Karger.

Block, W., de Jongh, J. G. V., Overbeek, J. T. G., and Sparnaay, M. J. 1960. *Trans. Faraday Soc.* **56**: 1597.

Bryson, G., and Bischoff, F. 1968. In *Progr. exp. Tumor Res.* **9**: ed. F. Homburger, p. 78. Basel/New York: Karger.

Derjaguin, B. V., Abrikossova, I. I., and Lifshitz, E. M. 1956. *Quart. Revs.* **10** (3): 295.

Gurwitsch, A. 1923a. *Arch. Entw.* **100:** 11.

———. 1923b. *Arch. Entw.* **103**: 490.

———. 1932. *Die Mitogenetische Strahlung*. Berlin: J. Springer.

———, and Gurwitsch, L. 1925. *Roux Archiv.* **105**: 470.

Harvey, E. V. 1940. *Living Light*. Princeton: Princeton University Press.

Metcalf, W. S., and Quickenden, T. I. 1967. *Nature* **216**: 169.

Protti, G. 1934. *Radiobiologia* **1**: 49.

Rahn, O. 1936. *Protoplasma-Monographien*, Vol. 9. Berlin: Verlag von Gebruder Borntraeger.
Selye, H., Priorischi, P., and Barath, M. 1961. *Oncologia* **14**: 65.
Troitskii, N. A., Konev, S. V., and Katibnikov, M. A. 1961. *Biofizika* **6** (2): 80.
Weisskopf, V. F. 1965. *Science* **149** (3681): 1181.
———. 1968. *Sc. American* **218** (5): 15–29.

CHAPTER 9

FURTHER IMPLICATIONS

Further predictive aspects, with positive implications for physiology, pharmacology, and pathology

Through quantum formulations physicists have arrived at a few rules which give the basis for the periodic classification of chemical elements and the origin of resonance or exchange energies in chemical bonds. It may be expected that an appropriate quantum formulation will be followed by a few rules governing many biological phenomena, as well.

Two such areas of general applicability may be realized. On the one hand, the negative guiding principle tells us how much can be known of the physical and chemical basis for a given biological process. We may be enjoined from giving a precise physicochemical description of a memory trace in the nervous system, for example. We may also find that the rate and specificity of an enzyme are quite indeterminate in the milieu of an organelle. We are under any circumstances prevented from taking the Cartesian view of the cell as a machine with a mechanistically determinate basis of operation. Conversely, a rapid perusal of the positive aspects shows us that we may find the time interval in which entropy does become determinate with obvious implications in aging, carcinogenesis, essential arterial hypertension, vessel sclerosis and thrombosis, and some forms of arthritis, as a few examples of pathological processes. Likewise, laws for epigenesis, extranuclear inheritance, and sexual conjugation may follow such an approach. We may hope for rules under which the storage of molecular memory may be subsumed in the antigen-antibody response, as an important instance. A more fundamental understanding of some of the phenomenological laws of physiology and pharmacology may be expected.

ARNDT-SCHULZ LAW AND THE LAW OF INITIAL VALUE AS IMPORTANT PHENOMENOLOGICAL LAWS OF ASYMMETRY OF METABOLISM

In Chapter 4 a reference was made to the Arndt-Schulz Law of pharmacology and Law of Initial Value of physiology as being important general instances of multivalued functions in metabolism which are analogous to the multivalued behavior of incremental negative resistance elements in electronics. This subject is discussed now in greater detail.

Metabolic processes require continuing energy input as do incremental negative resistance elements. In both, the extensive variables of energy are multivalued functions of intensive variables or vice versa. If plotted on a plane with coordinates to represent work (energy), the multivalued or discontinuous nature of these functions is apparent (see below, Figures 16A and 16B of voltage-current).

The Law of Initial Value has been explicitly connected with relaxation oscillation which requires negative resistance. The Arndt-Schulz Law has not hitherto been explicitly connected with negative resistance, but we point explicitly to the connection in this writing. This is reinforced by the general phenomenon of enhancement of single cell irritability before decrease in irritability under the influence of anesthetics, which Heilbrunn (1943, Chap. 27) repeatedly described. All fat soluble and "metabolic" anesthetic agents produce pre-anesthetic stimulation before anesthesia, an example of the universal tendency for poisonous substances in low concentrations to stimulate biological processes (the Arndt-Schulz Law) (Lamanna and Mallette, 1965). Numerous diverse, organic, and inorganic poisons have been investigated for their stimulating influence upon metabolic activities of many bacterial species, upon growth rates of various organisms, and upon crop yields. The stimulatory concentrations are slightly less than the minimal harmful concentration. The universal occurrence of stimulation by poisons has led many observers to consider the possibility of a single basic mechanism. However, no single deterministic mechanism has been identified or defined. We point to the Arndt-Schulz Law as a universal statement of metabolic phenomenology for which quantal explication is strongly suggested.

When this law of pharmacology is carefully reviewed with the very general phenomenological law of physiology known as Wilder's Law of Initial Value (Wilder, 1962; Block and Bridger, 1962; Brandt, 1962; and Mock, 1962), a most provocative likeness to the discontinuous or asymmetric or multivalued behavior of incremental negative resistance in electronics is recognized. The close connection of these two laws with temporal rhythms and homeostatic cycles of metabolism (living clocks or inherently open biological systems as Brown [1959] has described them) makes some single underlying basis of operation almost inevitable.

This underlying basis of operation is molecular indeterminacy of a nonconventionally statistical type which necessitates discontinuous or quantal explication.

The Law of Initial Value is an empirical statistical law for quantitative relation of response and stimulus in living systems; to quote Wilder (1962):

> Given a standard dose of stimulus and a standard period of measurement the response, that is, the change from the initial (prestimulus) level, will tend to be smaller when the initial value is higher; this applies to function-raising stimuli. For function-depressing or function-inhibiting stimuli this negative correlation becomes positive. Furthermore, beyond a certain medium range of initial values there is an increasing tendency *sometimes* toward absent and *frequently* toward reversed (paradoxic) responses; the higher these extreme initial values, the more frequent are the reversals on function-raising stimuli; the lower the initial values the more frequent are the reversals on function depressing stimuli.

Any experimentalist in the biological laboratory or practitioner in medicine will attest that the autonomic systems of various higher animals do not give identical responses to identical stimuli. Selbach (1962) may be quoted in this regard:

> Wilder's law of initial value (LIV) states that (1) the higher the state of stimulation is at the moment of the onset of stimulus, the weaker will be the additional excitation; thus, the response to stimuli with homodirectional inhibiting tendency will be weaker and the response

> to those with contradirectional inhibiting will be stronger; (2) if the initial value in the active system has already transgressed its medium action range prior to the onset of stimulus and has reached higher levels, a paradoxical reaction (reversal of effect) can occur.

Selbach ascribes paradoxical autonomic reactions as effects of the Law of Initial Value which have a triphasic course in systems with high unidirectional deviations from homeostasis and with low stability. He states that they can be analyzed according to the principle of relaxation oscillation as a special instance of the Law of Initial Value in cybernetic functions. We have come to a similar conclusion from an entirely different approach, which lends support to the fundamental nature of this subject.

One of our interests has been in attempting to find bona fide generalizations concerning phenomena of living systems in order to examine the conditions of these generalizations and to give a grounding in biological reality to our classes of transformations in a broadened application of quantum theory. These two laws fulfill a goal that we have been seeking: important generalizations for phenomenology of metabolism. These two generalizations are quite nicely covered under "incremental negative resistance" as a generic term which Mackay (1958, 1960) has advocated.

Mackay points out that a nerve can be considered a cascaded series of bi-stable elements. This does not explain the mechanism of functioning, but it does summarize many of the observed properties of nerves. The nerve impulse can be considered as a switching between two rather well defined states. The same is true of rapid metabolic events in muscle. As Mackay comments, whenever in nature a sudden switching action or bi-stability is observed, whether in living or nonliving processes, regenerative feedback is usually present. In an electrical system, the bi-stable function is associated with a negative resistance property. Without such a property in electric systems, direct current could not be converted into alternating current, nor could triggering or switching circuits be built. When an electrical circuit exhibits a negatively sloping relation between voltage input and current flow, that circuit contains the fundamentals for an oscillator, counter, or amplifier.

The chemist would be familiar with such multivalued or regenerative feedback systems under autocatalysis, and the thermodynamicist under activation energy. Mackay remarks that the social worker would use the term vicious cycle.

We are adopting the general concept of incremental negative resistance as a useful concept for our discussion and must now illustrate it with energy curves of voltage feedback and current feedback in electronic elements, each characterized by a negative slope. See Figures 16A and 16B.

ENERGY CURVES FOR INCREMENTAL NEGATIVE RESISTANCE ELEMENTS

If a steadily increasing voltage is applied in the energy input to certain electrical circuits or elements, the current will be seen to increase, then decrease, then increase in a discontinuous or multivalued fashion with respect to the voltage (Fig. 16A).

If a steadily increasing voltage is applied in the energy input to another type of electrical circuit or element, the current will increase a small amount, then suddenly jump to a higher flow, followed by a steady rise (Fig. 16B).

Evidence that incremental negative resistance processes are operating would be exhibited in any energy curve in which the extensive and intensive variables are representable by multivalued functions. The Law of Initial Value and Arndt-Schulz Law for metabolic processes can be illustrated as operating in the manner of the opposite knees of an energy curve with incremental negative resistance. (See Figure 17.)

If a steadily increasing physiological stimulus is applied in the energy input to a metabolic process, the physiological response may be seen to increase, then decrease, then increase. The discontinuous or multivalued behavior is described as the Law of Initial Value.

If a steadily decreasing amount of a pharmacological poison, beginning just above the minimal amount for poisoning, is applied in the energy input to a metabolic process, the pharmacological response may be seen to decrease, suddenly increase, and then decrease. The discontinuous or multivalued behavior is described as the Arndt-Schulz Law. This discontinuous behavior in induction of anesthesia of animals is a familiar phenomenon which is called the "excitement" stage by anesthesiologists.

It is quite satisfying that these acknowledged generalizations of a number of metabolic processes which fit the "incremental negative

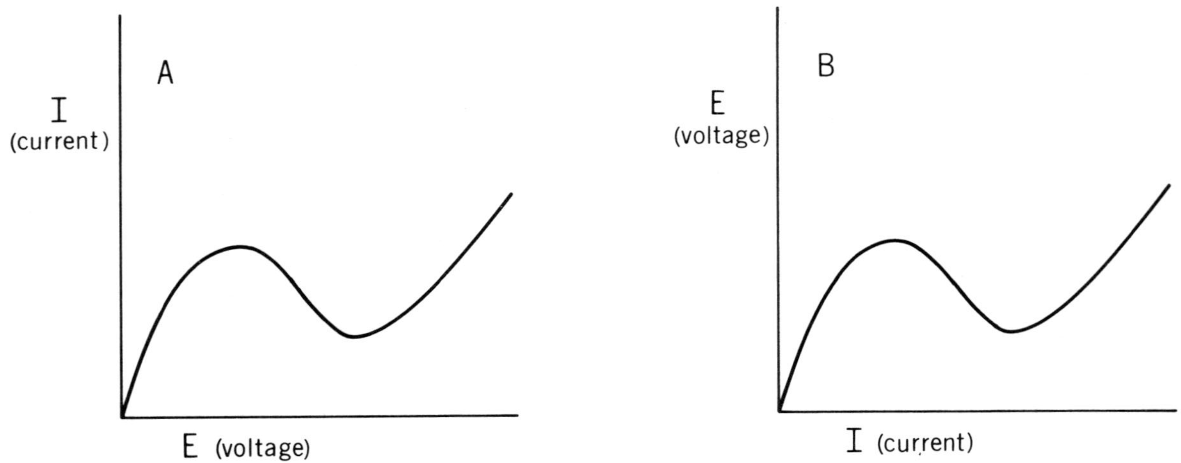

Figure 16. *A. Negative resistance as decrease in current for increase in voltage; B. Negative resistance as decrease in voltage for increase in current (from Mackay).*

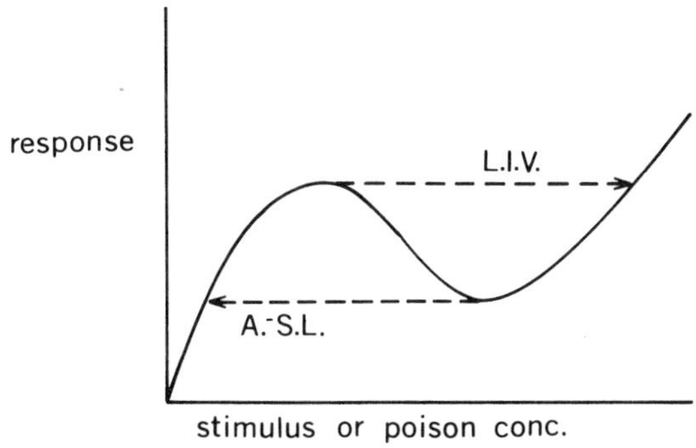

Figure 17. Opposite "knees" of a metabolic energy curve. The Law of Initial Value (L.I.V.) of physiology and Arndt-Schulz Law (A.-S.L.) of pharmacology appear as opposite "knees" of a metabolic energy curve of incremental negative resistance with response as a multivalued function of stimulus or poisoning concentration.

resistance" concept are described under the class of multivalued transformations which, on theoretical grounds, has evolved in this monograph for description of metabolic processes. These metabolic processes include nervous and muscular action (Johnson, Eyring, and Polissar, 1954) homeostasis (Cannon, 1932; Ashby, 1963, pp. 81–5 and Chap. 10), and temporal rhythms (Sollberger, 1962) and spatial patterns of mass and energy handling. (An outstanding and readily visualized example is Starling's law of the heart with its descending limb.)

HORMONES, ANTIGENS, HYPERSENSITIZING AGENTS, ALLERGIES

It was intimated in Chapter 3 that the baffling and inordinate effects of a seemingly insignificant number of molecules of agents such as hormones, antigens, allergens, and hypersensitizing materials would be explicable easily in terms of quantal effects, in keeping with the indeterminate basis of biological action. As a matter of emphasis, little alternative except quantal explication remains in the presence of the repeated observations that as few molecules as one per cell in an organism can initiate or prevent blood clotting, augment or inhibit muscle contractility, stimulate or suppress renal urinary secretion, activate or inactivate various aspects of liver metabolism, elicit or block hyperimmune reactions, and initiate or prevent depolarization of cell membranes. It was pointed out that if anything is overwhelming in modern endocrine research, it is the increasingly general observation that many different hormones are associated with the same particular moderating effect on an organism and that many different moderating effects are associated with the same hormone. Davidson (1965) has reviewed the influence of the same hormone on reactions both in the nucleus and in cytoplasm. These general observations with regard to "gene activation" and metabolic moderation by the same hormone should begin to weigh heavily upon causal explanation, which conversely loses impressiveness as further observations with hormones show these actions to be more qualified and interconnected than originally accepted. More important, an appropriate hypothesis for hormonal action is a by-product of recognizing molecular indeterminacy in many essential living processes. The hormone assumes a description as a bionic synthetic product which is involved in parameter selection of the major classes of transformations (genetic, epigenetic, metabolic) in other bions under which many living processes occur, *the whole operation having an indeterminate basis.* Antigens do the same, but are from alien sources, as are allergens and hypersensitizing agents. The parameter selection may vary greatly with the different agents and thereby exhibit different biological expressions.

ORGANISMIC LAWS AND SPECIFIC, LONG-RANGE FORCES IN BIOLOGY

One of the most critical consequences of uncertainty rules dealing with information handling can be predicted to be the general form which the organismic or emergent laws will assume. Here one can predict, for example, by close analogy with quantum physics, the existence of informational fields between large molecules. We have stated that such fields or directionalizing bonds would serve a very prominent role in the self-organization of both proteins and nucleic acids; in the noninvolvement of DNA directly in the replicative process (and subsequent permission for epigenetic processes); in the recognition occurring between proteins and between cell surfaces; in the specificity of antigen-antibody reaction; in the formation of filamentous, membranous, and granular organelles (many of which are really bions) and cells and the limitation of their sizes; in the requirement for the protein background in enzyme actions by (*a*) presenting a large surface for favorable collision, in order that high kT impacts can be utilized in the enzyme molecule to overcome activation energies, and (*b*) partitioning activation energies among catena, such as the electron transport particle; and in many other processes, even in the underlying orientational forces of muscle contraction, for instance. The forces constituting such information fields may be given a plausible description as the coupling of vibrational modes in these large macromolecules. We have described them as possibly electromagnetic in nature, albeit anomalous, resulting from the interaction of induced dipoles in the molecules of macromolecules. They are relatively weak of themselves—of the order of ten times thermal energies present in the medium, but additive because of the large numbers of participating modes. They have quite a long range resulting from the long wavelength associated with such vibrations, possibly on the order of a thousand angstroms to several microns. They are also very specific for the reason that these vibrations do assume very discrete values and the strength of coupling depends on the occurrence of like modes in the interacting molecule. They may be anomalous in distance-dependence because of coherence. These forces have been described as exchange of virtual photons between molecular oscillators under nonconventionally statistical conditions.

NUCLEAR AND EXTRA-NUCLEAR INHERITANCE

In line with these comments, the current theory of one gene representing one amino acid sequence for one enzyme protein as the form of information storage and as an information handling mechanism of the biological cell is in need of review from the standpoint of complementarity. The requirement of extra inducer genes, suppressor genes, and genes for redundancy to protect against noise is a great information burden to place on the chromosome.

In the light of complementarity, some of the information may be carried by cytoplasmic elements themselves. The proteins already carry part of their own code. The interaction of genes and proteins under non-Gaussian conditions can include a large amount of information for protein synthesis. Precise timing or phasing under quantum rules in the above interactions could perform some of the functions now ascribed to inducer and suppressor genes. The quantal aspects of this nonmechanistic interaction would effectively insulate the information moiety from thermal or entropic degradation while permitting the "dither" of thermal agitation to assist in the very synthetic process. Hinshelwood (1956) has pointed out the immense importance of cell dynamics to information handling apart from structuring of DNA's and RNA's. A more elegant complementarity than this between structure and function can hardly be found among the biological processes. Experimental observations which bring out "rigid" structural relations, as do x-ray diffraction studies of the structure of DNA and RNA molecules, destroy both the cell from which these molecules were taken and knowledge of functional information handling. Functional information studies on coding and genetics destroy or distort the cell or organism and do not give us knowledge of structuring. This is the principle of complementarity in essence, and it leads us to be wary that the quantal aspects of information processes are not overlooked. Attention is directed to our calculation with the DNA's in *E. coli* which casts strong doubt on complete determinability of information handling in the form of the classical "coding" systems currently being accredited. This is such an important point that it bears reiteration: genetic information handling in the biological cell requires the interactions of DNA's with the rest of the cell and the rest of the cell with DNA's. Being indeterminate from this viewpoint, genetic information is quantized. This means that, whereas the handling of the nucleotide sequence of the DNA's cannot be interpreted classically, it is not an improvement to say that the nucleotide sequences are quantized. Succinctly, the accent is that the interactions of the rest of the cell and the DNA's constitute genetic information, which is quantized.

There have been recent reports of the occurrence of patients with well-characterized clinical manifestations of trisomy–18 syndrome and trisomy–G syndrome (mongolism) but with normal karyotype, and a report of the familial occurrence of the phenotype of Turner's syndrome with normal karyotype (Hook and Tunis, 1965; Migeon and Whitehouse, 1967; and Hall, 1962). Such a dissociation of phenotype from a mechanistic genotype is extremely interesting in the light of our postulated information transfer processes.

Williams and associates (1962) have stressed that there is a high degree of inborn individuality even in inbred strains of organisms and that the basis for these inborn differences is presently out of reach of classical genetics and will be understood only when and if the extrachromosomal and other factors involved in differentiation are elucidated. Simple data like organ weights bear out this contention.

SEXUAL CONJUGATION

The necessity for "sexual" conjugation in cellular life forms is now seen as a sort of informational correction device. The breakdown of essential (dispersion coupled) information in an asexual cell line over several generations will be minimized or eliminated or reversed by required coupling with the large molecules of the cytpolasm of a new cell. In this case the life time of the cell (division time) is short compared to the rate of information degradation.

In sexual conjugation a distinct advantage is achieved in preventing error in information handling by one to one reproduction rather than one to two reproduction as in asexual fission or mitotic division.

INFORMATION QUANTIZATION: PROTECTION FROM "NOISE" WITH IMPLICATIONS FOR AGING

The inclusion of entropy in a complementarity relation allows any process which can be described in terms of information handling and also satisfies the uncertainty condition—product of the uncertainties of the same order of magnitude as the action—to be immune from noise to a much greater extent. Classical behavior implies, then, a direct and strict running down due to noise; on the contrary, behavior of a molecular process which is indeterminate on an unusual phenomenological basis confers a quantization of information levels in close analogy to quantum mechanics with quantized or somewhat immune energy levels.

Now suppose coupling of vibrational modes of macromolecules and electro-elastic collisions form an important part of cellular information processing in the form of the organismic laws of the cell. In this manner, even the nucleic acids may be pictured as having important informational coupling into the cell milieu. Over a considerable time interval this coupling may be subject to entropy increase due to noise, etc. Because of this, the ensemble now begins to form a time average of states or to "degenerate" to Gaussian numbers of operation. The interval of this shift toward thermodynamic equilibrium may be about the lifetime for the cell, and large chance increases of entropy may account for aging in cells existing for long periods in the life of the organism. Ruderfer's (1949) comments on the role of "action" in aging are interesting in this regard.

CARCINOGENESIS

Neoplasia can be viewed in a new light.

An important question to ask is whether a cell line propagates itself with an increase, over a long period of time, of the probability of losing its informational constraints.

This question seems to be answered in the affirmative by a consideration of several facts of biology.

Cyclic reproduction and sexual conjugation through exchange of material between germ cells of different organisms appear to be of general occurrence and appear to be important in maintaining vitality. Even bacterial organisms seem to have to conjugate every so often with other bacterial organisms to keep from losing vitality. Over many reproductions, this increase in probability of losing information would become prohibitive if sexual conjugation did not offer an information correctional and conservational capability.

More specialized animals set aside cells which do not undergo epigenetic division of bions. These primitive cells are germ cells which are involved in reproduction of the whole organism. Other cells very early take on specialized capabilities and are not ordinarily the cells for germination of whole new organisms. These "somatic" cells with epigenetic divisions of bions have emergent processes of information transport in which reciprocity has taken place with bions in other cells with formation of specialized tissues. These cells do not undergo sexual conjugation, but do undergo asexual mitosis or amitotic division for information conservation.

It appears that the existence of sexual conjugation and asexual mitosis and amitosis is interpretable as evidence that there is an increase in probability of information degradation with time, which these processes correct.

The simplest explanation for the sixth power law of increasing incidence of neoplasia in man with increasing age (Huxley, 1957) is a two-factor hypothesis of neoplasia. If the first factor of a two-factor hypothesis of neoplasia is increase in probability of losing information with time, the second factor can be any agent, agency, or process which enhances the rate of change of this increasing probability for information loss. The second factor might be high energy radiation, viruses, chemical carcinogens, or any of the large list that has been linked with increased incidence of neoplasia. These influences operate through alteration of the *emergent* metabolic processes which maintain fidelity of information transport of epigenesis or whole organism genesis (reproduction).

The time lag of a few months to 20–30 years or more between application of a known carcinogenic influence and the development of a tumor or working out of the carcinogenic process would be consistent with the operation of the second type of factor in enhancement of the first factor. Furthermore, the indeterminate basis of operation and emergent nature of the processes involved are attested by the failure to demonstrate a mechanistic basis even for high energy radiation-induced carcinogenesis despite great effort (Blum, 1959; Riley, 1966).

Some clarification concerning the second factor must be added. The fidelity of epigenetic and genetic information transport is maintained through emergent processes of metabolism. These emergent metabolic processes work in a manner that is similar to that of "incremental negative resistance" elements in electronics, as "Q" multipliers for tuned electronic circuits, or with bridges containing them, as "Q" meters (Mackay, 1958). This analogy is intended to carry through the accent of the multivalued nature of processes of metabolism. The point is that these metabolic processes are not just any metabolic processes. In these terms, the first factor of the two-factor hypothesis of neoplasia is the time-dependent increase in probability of genetic and eipgenetic information degradation, and the second factor is any influence which enhances the rate of this time-dependent information loss by alteration explicitly of the *emergent metabolic processes which maintain the fidelity of genetic and epigenetic information transport processes.* Warburg's (1956a, 1956b) classic finding of anaerobic metabolism with various cancerous tissues may be cited as an instance which illustrates the point. The finding of various anomalies of energy input in cancer cells points to a generalization that interference with energy input is a factor in neoplasia. This has apparently not been overtly recognized because of the paradoxical nature of metabolic energy input in the multivalued manner of negative incremental resistance.

This discussion and earlier chapters lead to the apparent fact that the occurrence of somatic mutation or spontaneous mutation is equivalent to a quantum leap to new informational states and is compatible with our thesis. Yet the question of whether mutation has occurred in a particular tumor is bound to be a confused and even an academic one, since in a population of tumors there are those which are not "caused," those which seem to be definitely linked to various definable influences, and those which may possibly be linked to various influences. The overlap of the two factors is so great that it may be impossible to make a positive distinction.

In view of the above discussion, an interesting differentiation in pathogenesis can be made. Tumors can be distinguished as instances in which in some manner those information transfer modes (transformations) that govern the division rate of cells are erratic. "Degenerative" processes appear to be distinguishable as instances where emergent epigenetic information transfer processes have been lost. Epigenesis implies cooperation between bions in cells at a distance (for specialization). Some of these emergent epigenetic processes, for instance, may be lost in nerve demyelinating diseases, in connective tissue degenerative diseases, as in some forms of arthritis, and in blood vessel sclerosis. Many congenital malformations are instances in which appropriate epigenetic information transport processes and/or parameter selections which are necessary for special development of the organism have failed or been subverted in embryogenesis.

Here are insights concerning the relation of congenital malformations, degenerative diseases, and neoplastic growths which, while distinguishable in certain aspects, are emergent processes that have indeterminate bases of operation which make their pathogenesis inseparable or easily confused. It also answers to some degree why one of the persistent theories of carcinogenesis has had its roots in congenital malformations and another has had its roots in degenerative processes.

Also, two great groups of animal neoplasms, carcinomas and sarcomas, are seen as the most frequent kinds of neoplasms for several reasons. They involve epigenetically differentiated tissues with most tenuous emergent metabolic processes for maintaining fidelity of information transfer. Since such cells can not undergo sexual conjugation, and in some instances can not undergo mitotic division as a means for information conservation, and since such cells are particularly susceptible to information degradation concerning their rate of division, they must pass on this lack in a carcinogenic process.

On the other hand, the wildest and most rapidly growing neoplasms tend to come from more primitive or less epigenetically differentiated cells and tissues or from the germ cells which do not undergo epigenetic change except late and to a lesser degree for the support of the germ cells in reproductive processes.

One can, in addition, comprehend the relative absence or reduced number of tumors in plants and in lower metazoan animals where epigenesis has not occurred or is very limited. Also, one can comprehend that lesser epigenetic change in plants and less specialized animals goes along with (*a*) the greater capability for regeneration of the total organism by any one of the individual, more primitive, pluripotent cells, (*b*) the greater cellular pleomorphism, (*c*) the lower malformation rate, and (*d*) the smaller number of hormones.

Returning to further support of the two-factor hypothesis of carcinogenesis (the factor of increase in probability of information loss with time and any factor which enhances this), overgrowth of antagonistic inhibition by hormone-dependent tumors is almost a requisite. Tumors which can be inhibited by removal of hormone necessary for their growth or by adding a suppressing hormone probably behave that way because they have a built-in lack of information. The growing and dividing tumor cells pass on this lack of information that would have kept them inhibited (if they were healthy they would have been inhibited and would have passed on the ability to be inhibited). This preferably may be termed difficultly reversible treption or "turning" of characteristics rather than mutation (Serra, 1964). Let us assume that a tumor depends on estrogen for high growth rate; if the ovaries are removed or androgen is given to the organism, we may expect inhibition of the tumor growth for a while. The tumor finally loses this inhibition related to the growth reducing androgen administration or to estrogen removal.

Information degradation is speeded by withholding hormone or adding inhibiting hormone, even though growth rate may have decreased. Notice that rate of information degradation is distinct from growth rate.

Most cancer chemotherapeutic agents utilized to date probably operate in the manner outlined above for hormones.

Immunity from attack by the host's defense processes along with the changes in stem line and ploidy also follow from inheritance of lack of information coupling in daughter tumor cells. Additional brief comments on host-tumor cell specificity have been made in Chapter 7 under our discussion of the phenomena of host-specificity.

Apparently spontaneous cures of well-documented cancers have been associated with massive viral infections, which we interpret as reversal of information degradation or information correction in the neoplastic cells (similar to normal conjugation). Nevertheless, some viruses can act as carcinogenic agents if the information background of a cell is sufficiently degraded or if the particular virus contributes to deterioration of the cell's information rather than improving it. This provides an approach to the paradox of viral causes and cures for various forms of neoplasia.

VICARIOUS CALCIFICATION AND INHIBITION OF *IN VIVO* CALCIFICATION

Howard and associates (1967) have made an important observation regarding inhibition of *in vitro* formation of calcium phosphate hydroxyapatite structure by a peptide inhibitor which is derived from whole human urine or serum. This is important toward understanding something of the nature of the clinically pathological process of *in vivo* vicarious calcification. It is also a lead toward interpreting the physiological inhibition of *in vivo* calcification.

These workers have demonstrated that supersaturated solutions of calcium and phosphate ions can be prevented from crystallizing by such a low concentration of peptide inhibitor that no direct involvement can be postulated for peptide blocking of the crystallization sites. The prevention of crystallization of $CaPO_4$ hydroxyapatite structure by a quantity of peptide which is 16,000 times smaller than the amount of calcium or phosphate ions forces the conclusion, in our opinion, that the incipient nucleation operation is an emergent or indeterminate process with distinctly quantal characteristics. After nucleation, the process does become thermodynamic.

Howard and associates have also demonstrated that one nanomole of the peptide inhibitor prevents the binding to cartilage of up to 1700 nanomoles of calcium ions. When tested in this fashion, more peptide is required to prevent apatite formation. Thus, crystallization in collagen is different from either crystallization of simple supersaturated Ca^{++} and phosphate solutions or crystallization of supersaturated Ca^{++} and phosphate solutions in the presence of peptide inhibitor. The collagen macromolecular nucleation sites are a fourth factor, and this system would be more amenable to treatment under long-range forces.

The whole matter, when viewed from ordinary chemical concepts, seems mysterious and bizarre and, in fact, can not be explained on the basis of stoichiometric chelation with free calcium ions. It is much clearer when looked upon as if it were a process of incipient nucleation which occurs when a few ions get together under conditions of micro-particulate motion concerning which conventionally statistical assumptions can not be made. It is cited as another intriguing biological observation which would fit nicely under the type of explication which is advanced in this monograph.

M CONCENTRATION OR SPACE IN BACTERIAL GROWTH

The bacteriologist witnesses a profound phenomenon in which a certain amount of physical space is required for bacterial growth. The total yield of bacteria per unit volume of fluid culture medium tends to be constant for a given bacterial species. Bail (1929) has called this population of bacteria the M concentration. Organisms may be removed by centrifugation at M concentration where growth and multiplication has ceased, and a new inoculum is followed by growth and reproduction. Exhaustion of food or inhibition by toxic products can not be invoked as causative factors in the bacterial growth cessation under these conditions. Although one may point to a possible critical minimal concentration of food per unit of surface or volume of organism for growth and multiplication, an interpretation that some profound aspect of "biological space" is involved in bacterial (bionic) growth and reproduction can not be escaped. Lamanna and Mallette (1965) have stated, "apparently some unsuspected factor is responsible for the phenomenon." We point to this as an example of a phenomenon where quantal explication is strongly suggested.

It seems quite possible that mutual inhibition by mitogenetic radiation from one bacterium to another at close distances is at work, or that influences of interbionic long-range forces operate on metabolic properties (permeability, etc.) among the bacteria.

INSTANCES OF APPARENTLY QUANTAL GROWTH

In Chapter 3 we cited the discontinuous growth of *Spirogyra* and the plant, crocus, as instances of apparently quantal biological processes (Thompson, 1942). Bose reported that the crocus grows in jerks of 0.002

millimeters about every 20 seconds, and Hofmeister described a similar type of observable process in *Spirogyra* (both quoted in Thompson).

A series of experiments would have to be designed to make certain that physical factors are not responsible for these effects, which seem to be quantized growth.

First, the possibility of ordinary cell division in a chain of *Spirogyra* or crocus cells must be ruled out.

Second, synchronization of cell growth from light effects or metabolism must be ruled out.

Third, synchronization by diffusable chemical substances along the *Spirogyra* chain or among crocus cells must be ruled out.

Fourth, some type of Hall effect must be considered.

The whole experiment might also be turned around, and the energy requirements for growth of *Spirogyra* or crocus determined calorimetrically, in order to ascertain that each incremental growth spurt is indeterminate.

THE GRAM STAIN IN BACTERIOLOGY

The Gram stain is of vital importance in bacteriology. Aside from the immense practical value of this triphenylmethane dye–iodine mordant–organic solvent technic for differentiating bacteria, it raises the question of why there is a correlation of one reaction type with so very many other bacterial functions? Lamanna and Mallette list at least thirty-five major correlations. They have also commented that rupture of Gram stained bacteria in the presence of nonpolar solvents, which prevents hydrolysis and enzymatic activity, is followed by immediate decolorization; this raises the suspicion that the Gram reaction may involve a peculiarity of morphological organization rather than a specific kind of Gram staining material. "At any rate, each proposed mechanism requires so many different specific properties for the system that a critic is led to challenge such mechanisms on the purely statistical basis of it being unlikely that all such unusual conditions can be satisfied" (Lamanna and Mallette, 1965).

The Gram stain situation is reminiscent of the many biological instances of protoplasmic, cellular, or tissue holism which are so well illustrated by hormone synthesis in the thyroid gland. Thyroid follicles must be intact for synthesis of thyroid hormone.

The Gram stain phenomenon prompts us to suggest that no particular molecular substrate will be found to be the key to the phenomenon, but that this phenomenon is yet another of the readily available instances which indicate molecular indeterminacy and unusual biological quantizations.

NERVOUS ACTION AS AN EXAMPLE OF QUANTIZATION AT HIGHER LEVELS OF ORGANIZATION

Upon viewing the structural hierarchy of a metazoan organism or higher plant, one would expect to find a tendency for the energy of interaction or numbers of events to become larger and hence amenable to "classical" description. At some level in the progression, molecule–macromolecular complex–organelle (bion)–cell–tissue–organ, quantal description would not be needed to predict adequately the behavior of a set of elements acting in concert. For example, given the input–output descriptions of a set of "machines" whose molecular transport processes are indeterminate, such as the chloride secreting cells of the stomach or the mechanical output of the muscle sarcomere, the integrated output of such elements can now be described in terms of the pH of the stomach contents or of the tension of the muscle fiber.

On the other hand, some tissues or organs may be specifically constructed so that the dynamic interconnections (or input–output) of the various elements in themselves must be indeterminate. This may be so, even though transformations providing a complete description of the emergent behavior of an individual element be known. This retroversion leads to quantized emergent behaviors within the tissue or organ itself and, indeed, may become the predominate mode of operation. The synaptic connections of the nervous system seem to be ideally constructed for the quantization of the molecular transport processes underlying the transmission of the nerve impulse from cell to cell. The synaptic space is typically very small (one micron) and events must transpire rapidly—with an order of magnitude of the arrival time of the inpulse. (The calculation at the end of Chapter 3 is directly relevant to this point.) The nervous system, having this obvious retroversion, provides a clear example of an "organ" displaying emergent behaviors based on the topology and dynamics of the connections of its parts. Quantization of the underlying molecular motion is seen to give rise to emergent behavior even at the higher integrative and control levels of an organism.

Several aspects of nervous function may be mentioned here which have direct bearing on the general problem of interconnection. McCulloch and Pitts (1943) have given the name "crummy neuron" to the probabilistic aspects of axonal transmission of impulses under a constant set of conditions. Considerable redundancy must be postulated in the form of parallel circuits in order to achieve reliable behavior in a given circuit. The nature of such an argument would be changed considerably by taking into account the emergent behavior of the synaptic connections. Stevens (1961), after a long series of experiments involving sensory perception, reached the conclusion that many aspects of these phenomena could best be explained by the existence of a neural quantum. Such a "smallest unit of activity" is not to be identified with the all or none response of the axon, according to him, but is to be found in the integrative function of the sensory system. Such an argument can be fitted into the context of emergent behavior arising from the retroversion occurring in the nervous system.

STABLE INTEGRATIVE MICRO-STATES IN THE NERVOUS SYSTEM

What may the existence of quantized emergent behavior at the synapse contribute to the modes of functioning of the nervous system? Quantization *per se* imparts stability (permanence in time and/or space) to processes. A large degree of specificity also accrues to such processes, especially when considerations of frequency and phase of energetic inputs (incoming and outgoing impulses of the nervous synapses) must be taken into account. Without specifying in detail the transformations describing synaptic function or their physical embodiment, two extremes of a continuum of neural-net function become apparent. Quantized emergent behavior may lead to the more or less permanent establishment of preferred neural circuits with a high degree of spatial and temporal specificity as one extreme. At the same time, since the set of neurons *and* their synaptic interconnections both contribute to nervous state-like behavior, it becomes impossible to specify completely the exact circuit involved with a particular engram. The engram becomes disassociated, as it were, from particular structures and pathways. Indeed, functionally separate pathways may occur over the same set of neurons, or conversely, structurally different circuits may transfer psychologically identical information. The overlap of spatial components in the circuits intermediate to the above cases can serve as the basis for the integrative and associative functions of the higher nervous centers. The description of such modes of functioning of neurons and interconnections assumes an integrative statelike character. They may be tentatively termed *micro-states* on the basis of (*a*) their involvement with a small number of elements, (*b*) their partial topological localizability, and (*c*) their stability.

What has just been described as functioning modes of these stable integrative microstates bears a striking resemblence to the aggregate field theory first proposed in detail by Lashley (1929, 1931). This type of theory was propounded to account for such psychophysical phenomena as the plasticity of behavior, learning of gestalts, and the lack of precise localization of complex learned behaviors. While successful in accounting for these and other psychophysical phenomena, the aggregate field fails to explain several others. Aspects of genetic neurology, with the inheritance of complex behavior and specific neural regeneration, are difficult to reconcile to a completely free (neurally detached) set of waves. Orderly maps of the periphery on the neocortex are also difficult to account for by field theory. Extensions of the concept of micro-states, having a necessary but somewhat plastic topological component, conceivably may offer a synthesis between the connectionistic and aggregate field theories of psychophysics and the mind-brain dichotomy.

Projection to the Psyche

Without overextending the foregoing argument, one further heuristic step leads to the origins of knowing and consciousness. Clusters of associative and storage micro-states may be pictured as interacting in concert to contribute to global states of behavior. Very tentatively, such global states are identified with the empirical phenomena of psychology to which the term "state" has also been applied in common usage. Sleeping state, waking state, states of awareness, and anxiety state are used to denominate distinct psychic sets of behavior sufficiently persistent to merit the term. Many of these conditions, of course, may not represent states in the sense in which the term is first used here. Such a decision should be capable of being made experientially. Here possible epistemological grounding has been given to the idea of psychic states. (Analogously, "states" in chemistry, i.e. molecules, were known by their empirical sets of transformations long before the quantum mechanical treatment of electrons in molecular orbitals allowed a theorectical description in terms of energy states.)

Having established, if only tentatively, a basis for the existence of global or psychic states of mental function; transitions between such states become focal in themselves. Possibly certain psychotic states are those from which transition is difficult. The psychiatric truism that "everybody is a little bit schizoid" then represents transitions with a high "transition probability" between a normal psychic state and a forbidden state. The full-blown psychotic individual has for some reason a low transition probability back to the normal states. It is tempting to describe manic-depression as two readily communicating abnormal states both of which are difficultly accessible to the normal, now forbidden state. Again, chronic anxiety may represent transitions analogous to the radiationless transitions to ground of molecular spectroscopy wherein there is a diffuse input from higher to lower brainstem instead of a single purposeful reaction such as fear, for example. Electroconvulsive shock and electro-anesthesia (also electrical defibrillation of the heart) may represent energy inputs that stimulate transitions between states.

A prominent example of integrative state-like behavior is that of the flight-fright excitation of the autonomic nervous system. A somewhat disguised complementarity argument has been waged for nearly a half century concerning whether one feels the emotion of fright which causes him to run or whether one begins to run from a recognized danger which causes him to feel the emotion of fright. Causality apparently cannot be assigned because the conjugate variables of molecular motion in the neurons and neuron connections, represented in emotion and behavior, have a complementary nature.

We have not in this brief excursion into the nervous system attempted to describe the working basis for the known mind-brain dualism. Nor have we attempted to pursue many other aspects of psychophysics which may have an indeterminate molecular basis: the glial-neuron interaction; the generation and effects of the DC electric field of the brain; mass transport within axons; toxic, drug, and metabolic effects; embryological specificity of connections; and regeneration, for example. Also, no claim is made that *all* of the properties of the higher nervous system are necessarily dependent on indeterminacy of molecular motion. We wish merely to point out that the almost unique extent of retroversion possible in the nervous system interconnections brings into consideration definite emergent properties with the attendant necessity to search for proper descriptive modes.

The discussions have been carried this far only to make an appropriate projection of the predictive power of generalized complementarity and broader quantum explication.

This monograph also must dwell to some extent upon the "negative" implications which arise in regard to the validity of statistical mechanical interpretation of living processes. Chapter 10 picks up the thread in this direction.

References

Ashby, W. R. 1963. *An Introduction to Cybernetics*. New York: Wiley.

Bail, O. 1929. *Ergebnisse exp. Populationsforschung. Zitatsh. Immun. Forsch.* **60**: 1.

Block, J. D., and Bridger, W. H. 1962. *Ann. N.Y. Acad. Sci.* **98**: 1229.

Blum, H. F. 1959. *Science* **130**: 1545.

Brandt, E. N., Jr. 1962. *Ann. N.Y. Acad. Sci.* **98**: 1291.

Brown, F. A. 1959. *Science* **130**: 1535.

Cannon, W. B. 1932. *Wisdom of the Body*. London: Norton.

Davidson, E. H. 1965. *Sci. Amer.* **212**: (6) 36.

Hall, B. 1962. Lancet **2**: 1026.

Heilbrunn, L. V. 1943. 2d ed. *An Outline of General Physiology*. Philadelphia: Saunders.

Hinshelwood, C. 1956. *Proc. Roy. Soc. London B* **146**: 155.

Hook, E. B., and Tunis, J. J. 1965. *J.A.M.A.* **193**: 840.

Howard, J. E. *et al.* 1967. *Johns Hopkins Med. J.* **120**: 119.

Huxley, J. S. 1957. *Biological Aspects of Cancer*. London: Allen and Unwin.

Johnson, F. H., Eyring, H., and Polissar, M. J. 1954. *The Kinetic Basis of Molecular Biology*. New York: Wiley.

Kandel, E. R., and Spencer, W. A. 1968. *Physiol. Revs.* **48**: 65.

Lamanna, C., and Mallette, M. F. 1965. *Basic Bacteriology*. Baltimore: Williams and Wilkins.

Lashley, K. S. 1929. *Brain Mechanisms and Intelligence*. Chicago: University of Chicago Press.

———. 1931. *Science* **73**: 245.

MacKay, R. S. 1958. *Am. J. Physics* **26**: 60.

———. 1960. *IRE Tran. Med. Elec.* **ME-7**: 94.

McCullouch, W. S., and Pitts, W. 1943. *Bull. Math. Biophys.* **5**: 115.

Migeon, B. R., and Whitehouse, D. 1967. *Johns Hopkins Med. J.* **120**: 78.

Mock, D. C., Jr. 1962. *Ann. N.Y. Acad. Sci.* **98**: 1296.

Pollard, Ernest, 1963. *J. Theoret. Biol.* **4**: 98–112.

Riley, V. 1966. *Science* **153**: 1657.

Ruderfer, M. 1949. *Science* **110**: 245.

Selbach, H. 1962. *Ann. N.Y. Acad. Sci.* **98**: 1221.

Serra, J. A. 1964. *J. Theoret. Biol.* **6**: 371.
Sollberger, A. 1962. *Ann. N.Y. Acad. Sci.* **98**: 757.
Stephens, S. S. 1961. *Science* **133**: 80.
Thompson, D. W. 1942. *On Growth and Form*. London: Cambridge University Press.
Warburg, O. 1956a *Science* **123**: 309.
———. 1956b. *Science* **124**: 269.
Wilder, J. 1962. *Ann. N.Y. Acad. Sci.* **98**: 1211.
Williams, R. J., Pelton, R. B., and Siegel, F. L. 1962. *Proc. Nat. Acad. Sci.* **48**: 1461.

CHAPTER 10

NEGATIVE IMPLICATIONS

Some negative implications of the quantum hypothesis for statistical mechanical interpretation of many essential living processes

The parallel might aptly be drawn between the situation in biology today and the situation in physics at an earlier date in this century. Physicists, including Planck, could not make an adequate explication of the experimental data from black body radiation, utilizing the classical physical laws of their day. Similarly, there were the negative experiments of Michelson and Morley, regarding the effect of a mythical ether upon the speed of light. Reappraisals of the theoretical foundations of physics were required before Planck and Einstein could adjust theory to experimental observations, from which the quantum of action and speed of light emerged as constants.

Today in biology, five direct instances of negative experiments and bodies of observation must be contended with in any classical mechanical interpretation of essential living processes.

a) We have the data from a huge number of genetic experiments dealing with chromosomal information transfer which Elsasser (1958) has pointed out is grossly inadequate to account physically for the total information requirements of the biological organism. This is true even though experimental work in genetics is incomplete. Enough has been done already to permit a projection of inadequacy for classical mechanistic interpretations of information handling.

b) Likewise, there is the work of Nathansohn (1900) with *Spirogyra*.

Spirogyra are seen to reproduce mitotically in water. Placed in 1 per cent ether solution, they reproduce amitotically. The amitotically divided *Spirogyra* then are placed back in water where they resume mitotic division. The question is how can *Spirogyra*, without benefit of the efficient mitotic process, divide the chromatin material exactly in half each time? Even worse, how can it divide all the different kinds of chromatin material exactly in half? Yet it later resumes the ability to reproduce mitotically! How has the "essential" genetic information been made up in the *Spirogyra* with less than half the chromatin material? And of the right kind? This akaryokinetic cell division is another type of negative experiment which challenges the classical mechanistic approach to information handling in biology. This is suggested by Child's (1907) studies with amitotic cell division in ovaries of sheep tapeworm proglottids. To quote him, ". . . cells with nuclei which have passed through a long series of amitosis are quite capable of exhibiting the characteristic phenomena of typical germ cells."

c) In the third instance there is the long, careful, and repeated work of the Chambers (1961) with intracellular water-soluble hydrogen ion indicators *in vivo*. They have demonstrated the radical and necessary asymmetry in metabolism between the nucleus and cytoplasm in the form of a decided difference in hydrogen ion concentration in many different kinds of living cells. They have also demonstrated that a wound to a living cell by micropuncture is followed by equilibration of the hydrogen ion concentration in nucleus and cytoplasm and a distinct fall toward greatly increased acidity in both. The usual absence of Brownian movement which characterizes living cellular cytoplasm and nucleoplasm is lost concurrently with the hydrogen ion equilibration at acid levels. With commencement of Brownian movement cell death ensues. These observations are paradoxical to an interpretation of random molecular motion, ordinary thermodynamic activity, and usual statistical mechanical operation in essential living processes.

d) Two reports by Marshak and Marshak (1954) and by Bracket (1954) concerning DNA studies cast doubt upon genetic information handling exclusively by DNA molecular mechanisms. The Marshaks studied echinoderm eggs (*Asterias forbesii*, *Arbacia punctulata*, and *Trysneustes ventricosus*) with Feulgen strain and uptake of radiothymidine. The pronucleus and the eggs themselves contain no thymidine that could not be accounted for by the somatic cells or polar bodies in the preparation. Lack of Feulgen stain and absence of radiothymidine uptake were two different forms of evidence that no DNA was present. It might be remarked that these eggs can be induced to develop parthenogenetically. The work was done to pursue the notion of maternal inheritance and to support the thesis that RNA carries genetic information and that DNA and/or its nucleotides act only as regulators. These experiments seem to us to be negative experiments in that genetic information transfer could not occur in completely mechanistic terms through DNA.

Bracket demonstrated that DNA diminishes rapidly and markedly at the end of the blastula stage in *Rana* embryogenesis. These "negative" studies and various radiotracer studies, such as those cited by Hoffman (1957), that show complete disappearance of RNA at times in cells, need greater exploration and attention.

e) In the fifth instance, P_{32} radiotracer studies show incorporation of atoms into DNA (Hoffman, 1957) in twice the amount that should be incorporated just for synthesis of new DNA molecules. If the P_{32} and other atoms go into old DNA which is supposed to be the template for replication of new DNA and RNA, questions are raised concerning a totally mechanistic account of genetic information by the DNA template.

PROBLEMS WITH COLLISION KINETICS AND LARGE MOLECULE SYNTHESIS AND INTERMEDIARY METABOLISM

Elementary collision kinetics have been applied by E. C. Pollard (1963) to processes of phage DNA and protein synthesis, messenger RNA combination with ribosomes, and glucose metabolism. Messenger RNA combination with ribosomes is not rapid enough to explain observed synthetic rates. The other processes also have observed synthetic rates too rapid or barely supportable by collision kinetics, if perfect collisions are required. The formation of a temporary complex with enzyme molecules, which is suggested to eliminate the requirement for perfect collisions, necessitates activation energy, the source of which is in grave doubt. If the energy of collision kinetics were satisfactory for a 100 fold faster maximal synthetic rate than observed, energy would conceivably be available for activation energy, but it is not. Activation energy requirement, in turn, necessitates that successive reaction be produced in an organized or "programmed" system, which begs the question deterministically. Also, if perfect collisions are required for the maximal synthetic rate to be supported under collision kinetics, the stability of the synthesized molecules is in question; that is, no consideration has been made of the fidelity of the synthetic process, since there is no activation energy barrier to prevent reversal of synthesis. Yet, the stability of DNA is one of the keys to biology according to Schrodinger (1945).

Pollard derived the number of molecules diffusing across a hemisphere of radius, r, to make contact with a specific synthetic region, R. In discussing this, he pointed out that his formula has two untenable assumptions: (*a*) The concentration in the immediate vicinity of the synthetic site is zero, nd (*b*) there are sufficient molecules present and under appropriate conditions to permit the idea of a concentration to have meaning. The importance of his contribution, however, is quite great in furthering a plausibility argument against explications which depend on utilization of the central limit theorem and assumption of Gaussian distributions. It is most interesting that Pollard suggests the paradox that *in vitro* synthesis of a bacterial cell would require design of very small reaction vessels, yet cells in excess of one micron size require some mechanism to guarantee collision. (See comments on limitation of sizes of bions to about the micron range, Chapters 5 and 8.)

BRIEF CRITIQUE ON CURRENT EXPERIMENTAL APPROACHES TO BIOLOGY

How many of the biological facts, of "biochemical" and "biophysical" nature relate truly to living things or to how their constituent parts operate? We have in mind events of short time and of molecular dimensions. Can the observed thermodynamic and kinetic behavior of an enzyme, as we characterize it *in vitro*, for example, be extrapolated through the inherent complexities of structure and function found in the cell? Many facts are gathered pell-mell about biological molecules without the slightest regard for their relevance to the protoplasm from which they were ripped. Indeed, little concern is paid to whether the molecule itself survived intact during the disruption. The complementarity relation warns us that the action of an enzyme may be a function of its isolation procedure. (Grisolia, 1964).

There are several criteria upon which such extrapolations must depend, and which are tacitly assumed when *in vitro* subassemblies are conceptually related to cell or organelle behavior. By explicitly stating some of these difficulties, as in Chapter 3 and in the pages below, we may begin to realize what formidable ground biochemistry and biophysics face in "explaining" life.

The first of these is the assumption that enzymes, at least those well characterized, do retain their substrate specificity in the milieu of the functioning cell. This assumption is not proven. Indeed some crystalized enzymes *in vitro* display different substrate specificities with a change in pH, for example. *A forteriori*, the almost unknown function of the rest of the protein molecule other than the "active site" may have very profound implications on the behavior of an enzyme functioning in an assemblage.

Very well, we say, let us determine all the specificities under a range of conditions: pH, oxidation-reduction potentials, salt concentrations, protein concentrations, surface adsorption; then, by rights we can, at least in principle, fathom the behavior of an enzyme. By patient synthesis, we may yet arrive at a physicochemical model. But here a more serious difficulty enters. What is meant by intercellular pH and E_h? Do we have "Gaussian" numbers of free protons and electrons to have such thermodynamic systems? What are the activities and not the concentrations of the important cations? What happens to a protein when adsorbed onto an aqueous-fatty acid interface? Some of the above concepts might have no meaning when applied to the small volume of the cell in which events must occur in a finite time. In other words, when we are not allowed to replace a time ensemble with a space ensemble (in the sense of Gibbs), we then may not use the concept of chemical potential upon which most if not all chemical thermodynamic arguments are based. For an excellent example see Feldman (1956) who discusses aspects of the use and abuse of pH measurements.

CORRESPONDENCE RULE AND STATISTICAL MECHANICS

This takes us back to the correspondence principle which requires that at some verge quantal mechanical explication will correspond nondualistically with classical mechanical explication. We have endeavored to put forth the positive advantages of quantum interpretation and negative drawbacks or limitations to statistical mechanical interpretation, but we hasten to point out again that the verge where nonconventially statistical quantum description of living processes does correspond with classical mechanical description is at those phenomenological conditions where statistical mechanics appropriately applies. The nonconventionally statistical quantum description is not without bounds.

COMPARISON OF HISTORICAL DILEMMA OF BIOLOGICAL AND PHYSICAL THEORY WITH DILEMMA OF CHEMICAL AND PHYSICAL THEORY

Setlow and Pollard have aptly summarized the historical dilemma of biological and physical theory and compared this dilemma with that of chemical theory and physical theory (1964).

Chemical characterization of elements, and laws of combination of elements, developed concurrently with physical laws of motion and electrostatics and spectra of elements. Problems came in chemistry with the periodic systematics of valence and atomic properties. Problems came in physics with explication of spectral radiation, an atomic property which did not align satisfactorily with the Rutherford-Bohr concept of the atom.

New ideas were demanded with the overwhelming complexities in valence doctrine of chemistry and with failure in physics of classical quantum theory to explain the helium atom, molecular structure, radioactivity, and other phenomena.

These were forthcoming for both chemistry and physics in quantum mechanics and the Pauli exclusion principle.

Similary, along with chemical and physical theory, biological theory has developed with its characterization of living things, the concept of the cell as the unit of biological systems, and organic evolution (paraphrased as, environment forces selection of form in living systems).

With the introduction of Mendelian genetics, however, many sticky problems have arisen in biology with respect to "leaky" genes, "sex-linked" genes, "lethal" genes, "weak" genes, etc., and similar problems have arisen in physics with the nature, production, and biological significance of macromolecules such as proteins and nucleic acids. Clearly, the processes of embryogenesis are still lacking explication, whether these explications be biological, chemical, or physical.

With the introduction of enzyme oriented intermediary metabolism, conflicts have become apparent in biology with respect to homeostasis, hormonal control, and irritability or sensitivity to environmental triggering and conditioning. Conflicts are apparent in biochemistry with the nonspecificity, control, induction, and moderation of control of enzymes.

As with chemistry and physics, new ideas are demanded for resolution of these many conflicts in biological theory.

In view of the grave difficulties at almost every turn in reaching physical explications of the essential processes of biology through the exclusive utilization of mechanistically determinate theory of physics and chemistry, the foremost question is why the warnings of Bohr (1934, 1958), Heisenberg (1962), and Weizsacker (1955) concerning complementarity and uncertainty have gone practically unheeded. Our effort is not to disregard these repeated question marks and conflicts in physical, chemical, and biological theory, but to face the dilemma constructively. This may be done by examination of the rather ubiquitous and obvious limitation of numbers of particles and nonequilibrium or radically open nature of energy and mass interchange with environment of nearly every crucial living process. These phenomenological conditions imply that the central limit theorem with Gaussian distributions of probabilites of conjugate variables of occupied energy states of molecular motion is not appropriately assumable in these processes, making the processes indeterminate and hence quantal with respect to molecular motion. Once the plausibility of this quantization is recognized, the exciting adventure is underway for progress in the new and more general science of biology and without rancor concerning mechanistically determinate explication in the many instances where it is appropriate.

References

Bohr, N. 1934. *Atomic Theory and the Description of Nature.* London: Cambridge University Press.

———. 1958. *Atomic Physics and Human Knowledge,* New York: Wiley.

Bracket, J. 1954. *Arch. Biol.* **65**: 1.

Chambers, E., and Chambers, E. 1961. *Explorations into the Nature of the Living Cell.* Cambridge: Harvard University Press.

Child, C. M. 1907a. *Biol. Bull.* **12**: 89.

———. 1907b. *Biol. Bull.* **12**: 177.

Elsasser, W. M. 1958. *The Physical Foundation of Biology.* New York: Pergamon.

Feldman, I. 1956. *Anal. Chem.* **28**: 1859.
Grisolia, S. 1964. *Phys. Rev.* **44**: 657.
Heisenberg, W. 1962. *Physics and Philosophy*, New York: Harper.
Hoffman, J. G. 1957. *The Life and Death of Cells*. Garden City, N.Y.: Doubleday.
Marshak, A., and Marshak, C. 1954. *Nature* **174**: 919.
Nathansohn, A. 1900. *Jahr., Wiss. Bot.* **35**: 1.
Pollard, E. C. 1963. *J. Theoret. Biol.* **4**: 98.
Schrodinger, E. 1945. *What is Life?* New York: Macmillan.
Setlow, R. B., and Pollard, E. C. 1964. *Molecular Biophysics*. Reading, Mass.: Addison-Wesley.
Weizsacker, C. F. von. 1955. *Naturewessenschaften* **42**: 521.

GLOSSARY

ADDITIVE SPECIFIC LONG-RANGE FORCES——Each of the adjectives suggests a specific feature arising from the somewhat unique generation of these forces by biological macromolecules. Additive refers to the increase in strength proportional to the number of like oscillators taking part in the interaction. Specific refers to the uniqueness of interaction between like or similar oscillators; while long-range refers to the distance and phase dependence of the interaction.

ASYMMETRY——The marked directionality of transport processes in biology. For example, an organism must expend energy to maintain the separation between itself and its environment. Mass, energy, charge, and momentum may be moved in these transport processes.

ATHEROGENESIS——Process of formation of atheroma in the walls of arteries. (Atheroma is a gruellike, fatty degeneration beneath the inner layer of an artery, often with collagen fibers, new connective tissue cells, phagocytes laden with fatty substances, and calcification.)

AUTOTROPHIC——Capable of self-nourishment; specifically, capable of using CO_2 or carbonates as a sole source of carbon and a simple inorganic nitrogen compound for metabolic synthesis.

BION——[See pages 85–8.]

BLUE GREEN ALGAE——Single unit of cellular microorganisms which contain chlorophyll masked by bluish green pigments, lack a condensed nucleus and chloroplasts, reproduce only by simple fission, and sometimes have been considered related to the bacteria. They comprise the class, Myxophyceae.

COACERVATE——An aggregate of colloidal droplets displaying some definite structure or arrangement.

COMPLEMENTARITY PRINCIPLE——[See pages 24–7.]

COLLAGEN-VASCULAR HYPERSENSITIVITY——Pathological reaction which involves ground substance and collagen material in blood vessel walls, thought to be associated with hyperimmune response to foreign agents (such as bacterial products) or to substances which ordinarily are healthy constituents of the organism showing the response.

CONJUGATE VARIABLES——Two related variables of motion of a dynamic system which, in conjunction with the laws of motion of the system, completely describe its state and its change of state in time. There are an infinite number of such variables, but the pairs nicely separating dynamic and kinematic variables are energy and time or momentum (linear or angular) and position.

COUPLING——An operation in which two or more machines are joined to form one machine. This operation can be symbolized in a very general way by a transformation which describes how the transformations are joined which represent each machine. What is symbolized is the operation of the whole machine through stipulation of the conditions or inputs for operation of each machine. Two additional considerations arise, however. First, for small machines (small action) unavoidable uncertainties occur with respect to noninterference of one machine with another, or others, in coupling. This implies "indeterminate" machines and "quantal feedback" between machines. Second, for the coupling of four or more machines (determinate or indeterminate), the interconnections between machines becomes very complex. Such complex, interconnected systems cannot be considered as a set of independent feedback circuits, but only as a whole with emergent or holistic behavior, which can be goal-seeking in very complicated patterns. (See Ashby, *An Introduction to Cybernetics*. New York: Wiley and Sons, 1963. p. 54.)

EMERGENT——Appearance of unexpected mode of behavior at a higher level of organization through a rearrangement of preexistent entities. In this monograph the term is restricted to include only those processes whose molecular motion is indeterminate.

EPIGENETIC——[See page 89.]

ESSENTIAL HYPERTENSION——A cardiovascular disease of unknown cause which is characterized by high arterial blood pressure.

GRAM STAINING——A method for differential staining of bacteria, utilizing gentian violet as a dye, Gram's solution as a mordant, and alcohol as a leaching agent. The result is that certain species retain the dye while others are decolorized. (Gram's solution contains one part iodine, two parts potassium iodide, and 300 parts water.)

HETEROTROPHIC—Requiring outside sources for nourishment; specifically, requiring complex organic compounds of carbon and nitrogen for metabolic synthesis: opposite to autotrophic.

INFORMATION——The "surprise value" or rarity of an event. This is the same meaning that the term has in information theory. Strictly defined, information is inversely proportional to the logarithm of the probability of the occurrence of the event. (See Chapter 3.)

LIESEGANG RING FORMATION——The general name for periodic precipitation or reaction occurring upon the diffusion of a molecule or ion into a matrix containing the precipitant or reactant. Usually the matrix is a colloid, but not necessarily. The periodic formation in an agar gel containing dilute potassium dichromate of a silver chromate precipitate, when a drop of dilute silver nitrate is applied to the surface of the gel, is an example of Liesegang ring formation.

LIMIT THEOREM—A mathematical theorem in probability theory which makes a statement concerning the proper abstraction to use in characterizing a collection of a large number of probability variables of a certain type.

MECHANISTICALLY DETERMINATE PROCESS——A process in which the conjugate variables of motion (position and momentum, time and energy, etc.) can be measured simultaneously without interference with knowledge of the process. A complete mechanistic description can be made of such a process, in which prediction and postdiction of the trajectory of the process are possible.

MECHANISTICALLY INDETERMINATE PROCESS——A process in which the conjugate variables of motion can not be measured simultaneously without interference with knowledge of the process. The mechanistic descriptions that can be made of such a process are incomplete, probabilistic, and complementary. Prediction and postdiction of the trajectory of the process are not possible.

METALANGUAGE——A language of a higher logical type than a given system or discipline which can deal critically with the nature, structure, or behavior of the given system or discipline.

MITOGENETIC (MITOGENIC) RADIATION——A type of radiation given off from active tissues; the primary effect is on the rate of mitosis of nearby organisms.

M-SPACE OF BACTERIA——[See page 151.]

MYCOPLASMATALES——An order of microorganisms (coextensive with the family Mycoplasmataceae) of pleomorphic, Gram negative, nonmotile nature that are intermediate in some respects between viruses and bacteria, have complex life cycles, have no cell walls, and are usually parasitic, often in mammals.

NEGATIVE (INCREMENTAL) RESISTANCE ELEMENT——A term from the field of electronics designating a circuit element which over some range of its operation displays decreasing current with increasing impressed voltage (or vice versa). Such elements form the basis for many switches and other bistable devices, oscillators, and amplifiers.

NONDUALISTIC THEORY——A theory which while general enough to allow description of classical mechanical, statistical mechanical, ordinary quantum mechanical, and quantum statistical mechanical processes is sufficiently general to allow description of quantum mechanical processes of a broader nature, such as those whose phenomenological conditions do not permit assumption of Gaussian distributions of complementary variables in an indeterminacy relation. It is implied that under a correspondence rule there is a limit to the latter type of description as the boundary situation is approached in which phenomenological conditions do permit assumption of Gaussian distributions. In this case, the description now corresponds with statistical mechanical description, and a correspondence rule is obeyed. This qualifies the theoretical foundation of the description as nondualistic with that of physical theory.

ONCOGENIC——Giving rise to a tumor.

OPERAND——The element which is changed to another in a transformation.

OPERATOR——The factor, in cybernetics, which in acting brings about the transition between two elements of a system. Physical causality need not be implied.

PARAMETER——Signifies a means of selecting and performing an operation on a subset of available operands, when used in regard to a transformation.

PHASE SPACE——A multidimensional space of which the coordinate dimensions represent the variables required to specify the phase or state of a system.

PHENOMENOLOGICAL——Refers to observable facts or events of experience.

PLASTIC-ELASTIC (NATURE OF THE ACTIVE SITE OF AN ENZYME)——The central idea of Koshland's induced fit hypothesis of enzyme action. In this theory the secondary and tertiary structure of an enzyme in the vicinity of the active site is deformed by interaction with the substrate.

PLURIPOTENT——Refers to a cell's having the capacity to give rise to more than one type of cell in its later development.

RADICALLY OPEN PROCESS——Refers to a process which is exchanging mass or energy with its environment at such a rate that it may not be described as a series of approximations to a stationary state. These processes, when quantized, are asymmetric, multivalued in input or output or both, and discontinuous.

RADICALLY RECURRENT PROCESS——An inherently cyclic process, capable of description, but whose constituent molecular motion is severely limited in time or space or both. As a consequence, in these processes that do act quantally, information is conserved from cycle to cycle.

RICKETTSIA——Common name for organisms of the order Rickettsiales which resemble bacteria but typically live within the cells of a host organism. (See Chapter 5.)

STATIONARY STATE——The state of a system which does not vary with time. For quantized systems, the wave function, ψ, is not time-dependent.

TELEONOMIC——Refers to the programmed goal-seeking behavior of an organism or system. This is a very different idea from the end-directed casuality implied by the term teleologic.

TRANSFORM——The element to which another is changed in a transformation.

TRANSFORMATION——The most general method for representing the change between identifiable states of a system. The representation consists of elements known as *operands* upon which an *operator* acts to give rise to a set of elements known as *transforms*.

VICARIOUS CALCIFICATION——Calcification occurring in an abnormal situation in a biological organism.

VIRTUAL PHOTON——The carrier of the electromagnetic force between oscillating or otherwise accelerated charges, that is, photons which through their exchange between static electric charges and between currents give rise to the coulomb and magnetic forces respectively. Not "seen" as radiation since they are continually absorbed and re-emitted in interaction.

INDEX

THE JOHNS HOPKINS PRESS

Designed by Arlene J. Sheer

Composed in Monotype Baskerville with Baskerville Roman display
by Monotype Composition Company

Printed offset on 60 lb. Perkins and Squier R
by Universal Lithographers, Inc.

Bound in Columbia Fictionette
by L. H. Jenkins, Inc.